THE QUANTUM THEORY OF THE ATOM

THE QUANTUM THEORY OF THE ATOM

by

GEORGE BIRTWISTLE,
FELLOW OF PEMBROKE COLLEGE,
CAMBRIDGE

CAMBRIDGE
AT THE UNIVERSITY PRESS
1926

CAMBRIDGE
UNIVERSITY PRESS

University Printing House, Cambridge CB2 8BS, United Kingdom

Cambridge University Press is part of the University of Cambridge.

It furthers the University's mission by disseminating knowledge in the pursuit of education, learning and research at the highest international levels of excellence.

www.cambridge.org
Information on this title: www.cambridge.org/9781107463912

First published 1926
First paperback edition 2014

A catalogue record for this publication is available from the British Library

ISBN 978-1-107-46391-2 Paperback

PREFACE

THIS book opens with an historical account of the quantum theory from its inception by Planck, which is followed by a full treatment of the dynamical theory evolved by Bohr and his school during the past ten years to explain the phenomena of line spectra. The theory ranges from the earliest simple conception of the circular orbit of hydrogen to the application to the atom of the theory of perturbations of celestial mechanics.

Some chapters are devoted to a general description of optical and X-ray spectra and their significance in the problem of the atomic structure of the elements.

The book closes with an account of the new work on the reaction of the atom to radiation fields, which promises to lead to far-reaching developments of the quantum theory.

G. B.

CAMBRIDGE

Oct. 10, 1925

CONTENTS

THE QUANTUM THEORY OF THE ATOM

CHAPTER I

THE ORIGIN OF THE QUANTUM THEORY; PLANCK'S RADIATION FORMULA

1. *The origin of the quantum theory.* During the last fifteen years of the nineteenth century, the problem of the distribution of energy in the spectrum of a black body had actively exercised the minds of the ablest mathematical physicists, and had resisted all attempts to find a theoretical formula, based upon the mechanical principles of Newton and the electrodynamical theory of Maxwell, which would agree with the known experimental evidence.

The difficulties culminated in the proof by Jeans[1] in 1909 of a formula, established for all wave lengths, which was generally accepted as the only result to which the laws of Newton and Maxwell by themselves could lead. This formula was flatly in conflict with the results of experiment.

But for some time before this decisive result had been obtained, it had become more and more apparent that the classical theory (of Newton and Maxwell) was incapable, alone, of accounting for the absorption and emission of radiation by atoms; some new axiom was required.

The new theory was given in 1900 by Planck[2] who had devoted himself for many years to the application of thermodynamical principles to electrodynamics; in the course of this work the need for some form of *discontinuity* in atomic processes had become apparent.

He supposed that the oscillating electrons of the radiating

[1] J. H. Jeans, Phil. Mag. **17**, p. 229, 1909.

[2] M. Planck, Verhandl. d. Deutschen Phys. Ges. **2**, p. 202, 1900.

body do not radiate or absorb energy continuously, as required by the ordinary electrodynamics, but on the contrary radiate and absorb discontinuously. The energy of the 'oscillator' (oscillating electron) was supposed to be a whole multiple of a 'quantum' of energy of magnitude $h\nu$, where h is a constant (Planck's constant) and ν is the frequency of the oscillator (the number of oscillations per second). Thus the oscillator can only emit or absorb energy in whole multiples of the quantum $h\nu$; or, interchanges of energy between matter and radiation can only occur discontinuously in quanta.

The new theory, 'the quantum theory,' led to a formula for the distribution of energy in the spectrum in remarkable agreement with the results of experiment.

The quantum theory soon led to new advances. The first was due to Einstein, who readily explained the photo-electric effect[1], which had hitherto defied explanation in terms of the classical theory, and also found a formula which accounted for the decrease of the specific heat of solid bodies with falling temperature[2]. Einstein's view, which depended upon the assumption of a single free period for the atoms, was extended by Born and Karman[3] to the case of several free periods; while Debye[4] by a simplification of the assumptions as to the nature of the free periods found a formula for the specific heat of solid bodies which more closely represents its values, especially for the low temperatures obtained by Nernst[5] and his pupils, and which was in striking agreement with the elastic and optical properties of such bodies.

Again, Nernst[6] pointed out that to an energy quantum of vibration, such as that of Planck, there must correspond an energy quantum of rotation, and that it was to be expected that the rotational energy of gas molecules would vanish at

[1] A. Einstein, Ann. d. Phys. **17**, p. 132, 1905.
[2] A. Einstein, Ann. d. Phys. **22**, p. 180, 1907.
[3] M. Born and Th. von Karman, Phys. Zeitschr. **14**, p. 15, 1913.
[4] P. Debye, Ann. d. Phys. **39**, p. 789, 1912.
[5] W. Nernst, Zeitschr. für Elektrochem. p. 818, 1911.
[6] W. Nernst, Phys. Zeitschr. **13**, p. 1064, 1912.

very low temperatures. This conclusion was verified by Eucken's[1] measurements of the specific heat of hydrogen.

These and other results obtained in the most diverse regions of physics formed in themselves almost overwhelming proof of the existence of the quantum of energy, but it was not until 1913, when Niels Bohr[2] with brilliant success derived the Balmer formula for the spectral lines of hydrogen, and the Rydberg constant in terms of known physical magnitudes, that the quantum theory met with real recognition.

2. *Radiation.* Electromagnetic waves are known of a wide range of wave length—electric waves, infra-red rays, light, ultra-violet rays, X-rays, and γ-rays—a sequence of wave lengths diminishing from hundreds of metres for electric waves to lengths of order 10^{-10} cm. for γ rays. Denoting by Å., the Ångström unit of 10^{-8} cm. and by X, the X-ray unit of 10^{-11} cm., the visible spectrum extends from about wave length $\lambda = 7000$ Å. (red) to $\lambda = 4000$ Å. (violet). The longer infra-red rays have been observed up to lengths of order 100μ, where $\mu = 1000$ Å.; the shorter ultra-violet rays have been observed down to a few hundred Å. and these are the same as the very 'soft' X-rays of the lighter elements; the sequence of X-rays continues with decreasing wave length until the 'hard' X-rays of the heavier elements merge into the γ-rays. The ordinary range of X-rays is from $10{,}000X$ to $2000X$; the γ-rays have been observed down to $70X$.

If a body is kept at a given temperature and steadily emits radiation without structural change the radiation is 'pure temperature radiation.' Lamp black is a substance which can emit or absorb almost all wave lengths of temperature radiation; an ideal body which can emit or absorb all wave lengths is called a 'black body.'

Consider a hollow metal globe with a small hole in it. Any radiation entering the hole from outside would be partly absorbed by the walls and partly reflected. The chance of the reflected radiation ever emerging from the hole is very small

[1] A. Eucken, Sitzungsber. d. Preuss. Akad. Wiss. p. 141, 1912.
[2] N. Bohr, Phil. Mag. **26**, pp. 1, 476, 857, 1913.

indeed, and such minute portion as did so would have undergone so many reflections that its intensity would be very small. It may therefore be said that a small hole in a hollow body is a perfect absorber of heat, that is, is the experimental realisation of the perfect black body.

Hence if a small hole be made in the wall of a hollow metal vessel heated to some given temperature, the radiation from the hole will be that of the 'ideal' black body at that temperature.

By the use of a prism of fluorspar 'black-body' radiation can be dispersed into a spectrum. If the energy of the part of the spectrum corresponding to wave lengths between λ and $\lambda + d\lambda$, where $d\lambda$ is a small definite range of wave length, is measured by a bolometer and is $E_\lambda \,.\, d\lambda$, E_λ can be plotted as ordinate with λ as abscissa for all the wave lengths of the spectrum. The distribution of energy amongst the different wave lengths is thus known by experiment. The form of this curve was determined by the experiments of Lummer and Pringsheim[1] for a range of temperatures 620° to 1646° (Thomson degrees).

The form of the curve was

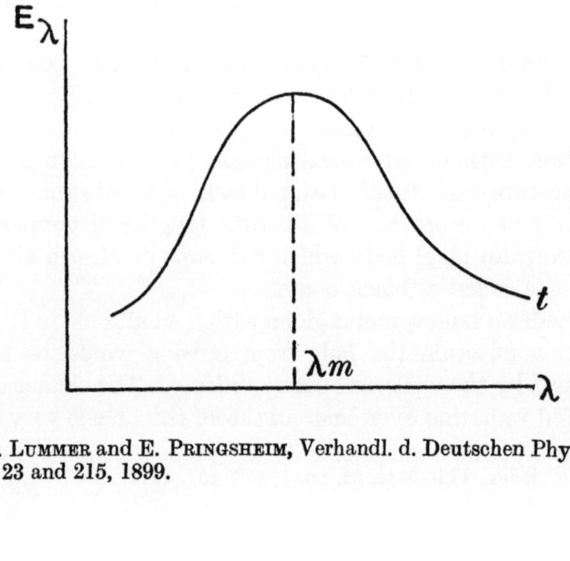

[1] O. Lummer and E. Pringsheim, Verhandl. d. Deutschen Phys. Ges. 1, pp. 23 and 215, 1899.

There is a wave length (λ_m) for which E_λ is a maximum at any given temperature.

3. *Wien's law.* The researches of Kirchhoff[1] on the emission and absorption of radiation, of Boltzmann[2] and of Wien[3], who applied the general principles of thermodynamics to radiation, led to the theoretical deduction of Wien's law that $E_\lambda = \frac{1}{\lambda^5} f(\lambda t)$, where f is a function entirely independent of the nature of the particular substance emitting or absorbing the radiation, and t is the temperature.

The maximum of the experimental curve would correspond to the equation

$$\frac{d}{d\lambda}\left[\frac{1}{\lambda^5} f(\lambda t)\right] = 0,$$

or

$$\lambda t f'(\lambda t) = 5 f(\lambda t).$$

This is a numerical equation for λt, and if c is the root, $\lambda_m t = c$; thus as the temperature rises λ_m diminishes, or, the maximum moves to shorter and shorter wave lengths as t increases. This is the 'displacement law' of Wien. The constant c was found by Lummer and Pringsheim to be $\cdot 294$ cm.-deg.

For the sun, Langley found $\lambda_m = 5000\,\text{Å}. = 5 \times 10^{-5}$ cm., so that

$$t = \frac{\cdot 294}{5 \times 10^{-5}} = 5880; \text{ or, } t \text{ is about } 6000°.$$

For the high temperatures in the interior of the stars of the order of 10^6 degrees, the corresponding λ_m would be of order $\cdot 294 \times 10^{-6}$ or 3×10^{-8} or $3000X$ and the maximum of the energy of the radiation would be of an average X-ray type.

[1] G. KIRCHHOFF, 'Gesammelte Abhandlungen,' Leipzig, p. 597, 1882.

[2] L. BOLTZMANN, Ann. Phys. Chem. **22**, p. 291, 1884.

[3] W. WIEN, Sitzungsber. Akad. Wiss. Berlin, p. 55, 1893. A full account of the deduction of this law is given by M. PLANCK in his book 'Wärmestrahlung' (fünfte Auflage), Leipzig, pp. 59–88, 1923, and by H. A. LORENTZ in his 'Theory of Electrons' (English translation). chap. II, 1909. Also 'Principles of Thermodynamics' by G. BIRTWISTLE, pp. 152–157, Cambridge, 1925.

4. *The function f.* It has been seen that the principles of thermodynamics lead to the formula $E_\lambda = \lambda^{-5} f(\lambda t)$, but they do not suffice to determine the analytical form of the function f. Further progress can only be made by the use of some hypothesis as to the nature of the emission and absorption of energy by the vibrating electrons ('oscillators').

The earliest attempt to deduce a form for f was published by Wien[1] in 1896. He assumed that the length λ of the waves emitted by a molecule is a function of its velocity v and that E_λ is proportional to the number of molecules having velocities between v and $v + dv$. This led to an exponential form for the function f.

Wien found that $f(x) = ae^{-\frac{b}{x}}$, so that

$$E_\lambda = \lambda^{-5} f(\lambda t) = a\lambda^{-5} e^{-b/\lambda t},$$

where a, b are constants.

In 1899 Lummer and Pringsheim (loc. cit.) had tested this formula by experiment, and in 1900 Rubens and Kurlbaum[2], using infra-red rays of much greater wave length, extended the test. Wien's formula was found to hold for values of λt below a certain value, but to fail for long wave lengths.

At about this time, Rayleigh[3] found the form $f(x) = cx$ for large wave lengths, where c is a constant, so that

$$E_\lambda = \lambda^{-5} f(\lambda t) = ct/\lambda^4,$$

for large wave lengths. This formula agreed well with experiment for long wave lengths.

Thus at this time (1900) there were two formulae,

$$E_\lambda = a\lambda^{-5} e^{-b/\lambda t} \text{ (Wien) and } E_\lambda = ct\lambda^{-4} \text{ (Rayleigh)},$$

the former holding for small wave lengths and the latter for large wave lengths, each in agreement with experiment for a certain range of λ.

5. *The discovery of Planck's formula.* For some years before 1900, Planck had endeavoured to determine f by some

[1] W. Wien, Ann. d. Phys. **58**, p. 662, 1896.

[2] H. Rubens and F. Kurlbaum, Sitzungsber. Akad. Wiss. Berlin, p. 929, 1900.

[3] Rayleigh, Phil. Mag. **49**, p. 539, 1900; 'Coll. Papers,' vol. **4**, p. 483.

new assumption as to the nature of the interaction between the 'oscillators' and the radiation. He considered the possibility of the oscillator being able to exert an 'irreversible' action on the radiation, to the consternation of his thermodynamical mentor Boltzmann[1]. In the course of his work he had used a function S, the reciprocal of $\frac{d^2\phi}{dE^2}$, where ϕ is the entropy and E the energy, S having a direct physical significance in connection with the irreversibility of the energy exchanges between the oscillators and the radiation.

He calculated S for each of the formulae of Wien and Rayleigh. Wien's formula is

$$E = \alpha e^{-\beta/t}, \text{ where } \alpha = a\lambda^{-5},\ \beta = b\lambda^{-1}.$$

But $$d\phi = \frac{dE}{t}, \quad \therefore \frac{d\phi}{dE} = \frac{1}{t} \text{ and } \frac{d^2\phi}{dE^2} = -\frac{1}{t^2}\frac{dt}{dE}.$$

But $$\log E = -\frac{\beta}{t} + \text{constant}.$$

$$\therefore \frac{dE}{E} = \frac{\beta dt}{t^2},$$

so that $$\frac{dt}{dE} = \frac{t^2}{E\beta},$$

$$\therefore \frac{d^2\phi}{dE^2} = -\frac{1}{E\beta},$$

$$\therefore S = 1\bigg/\frac{d^2\phi}{dE^2} = -E\beta.$$

Rayleigh's formula is $E = \gamma t$, where $\gamma = c\lambda^{-4}$,

$$\therefore \frac{dt}{dE} = \frac{1}{\gamma},$$

and $$\frac{d^2\phi}{dE^2} = -\frac{1}{t^2}\frac{dt}{dE} = -\frac{1}{\gamma t^2} = -\frac{\gamma}{E^2},$$

$$\therefore S = -E^2/\gamma.$$

Thus the function S for small wave lengths was proportional to E and for large wave lengths to E^2. Planck therefore tried

[1] L. BOLTZMANN, Sitzungsber. d. Preuss. Akad. Wiss. p. 455, 1899.

$S = -\beta E - \frac{E^2}{\gamma}$ as a possibility in general, the first term being important for small energies and the second for large energies.

This assumption led to

$$\frac{d^2\phi}{dE^2} = \frac{1}{S} = -\frac{\gamma}{\beta\gamma E + E^2}$$

$$= \frac{1}{\beta}\left[\frac{1}{\beta\gamma + E} - \frac{1}{E}\right],$$

$$\therefore \quad \frac{d\phi}{dE} = \frac{1}{\beta}\log\left(\frac{\beta\gamma + E}{E}\right).$$

Also since $$\frac{1}{t} = \frac{d\phi}{dE},$$

$$\frac{\beta}{t} = \log\left(\frac{\beta\gamma + E}{E}\right),$$

$$\therefore \quad \frac{\beta\gamma}{E} + 1 = e^{\frac{\beta}{t}},$$

or $$E = \frac{\beta\gamma}{e^{\frac{\beta}{t}} - 1}.$$

Inserting the values of β, γ, namely, $\beta = b\lambda^{-1}$, $\gamma = c\lambda^{-4}$,

$$E_\lambda = \frac{bc\lambda^{-5}}{e^{b/\lambda t} - 1},$$

where b, c are constants.

This new radiation formula[1], which was so far 'only an interpolation formula found by happy guesswork,' to use Planck's own phrase, was found to agree closely with experiment for all values of λ.

For large values of λt, it becomes

$$E_\lambda = \frac{bc\lambda^{-5}}{\frac{b}{\lambda t}} = \frac{ct}{\lambda^4},$$

the formula of Rayleigh, and for small values of λt, it becomes

$$E_\lambda = \frac{bc\lambda^{-5}}{e^{\frac{b}{\lambda t}}} = \frac{bc}{\lambda^5}e^{-\frac{b}{\lambda t}},$$

the formula of Wien.

[1] M. PLANCK, Verhandl. d. Deutschen Phys. Ges. p. 202, 1900.

From this time onward Planck was occupied with the task of giving this formula a real physical basis; he turned to Boltzmann's line of thought that entropy is a measure of physical probability.

6. *Entropy and probability.* In the thermodynamical theory of a gas, the state of the gas is defined by a small number of parameters, such as the temperature and pressure. In the kinetic theory of a gas, the state of the gas is defined by a large number of parameters, the coordinates and velocities of its molecules. The temperature and pressure are the mean values of certain functions of the parameters of the kinetic theory, so that a given thermodynamical state may correspond to a large number of different kinetic states ('complexions' as Gibbs termed them) for which the aforesaid mean values are the same. Boltzmann considered that a given thermodynamical state was the more likely to occur the greater the number of 'complexions' by which it could be realised, and he took this number (W) to be the measure of the probability of its occurrence.

7. *The equation* $\phi = R \log W$. Let ϕ be the entropy of a system in some given state and W the probability of its occurrence. Boltzmann took the view that Clausius' principle that a system tends to pass towards states of greater entropy is also the principle that systems tend to pass towards states of greater probability of occurrence, and wrote $\phi = f(W)$.

Consider two independent systems in respective states 1 and 2 for which the probabilities are W_1, W_2. Then if the two are considered as forming one system, the probability W for the system is equal to $W_1 W_2$, as any pair of 'complexions' of the separate systems taken together forms a 'complexion' of the whole.

But the entropy ϕ of the system is $\phi_1 + \phi_2$.

Therefore since $\phi_1 = f(W_1)$, $\phi_2 = f(W_2)$ and $\phi = f(W)$,

$$f(W_1) + f(W_2) = f(W) = f(W_1 W_2) \quad \ldots\ldots(1).$$

Differentiating both sides with respect to W_1, keeping W_2

constant, it follows that $f'(W_1) = W_2 f'(W_1 W_2)$, and differentiating this with respect to W_2, keeping W_1 constant,

$$0 = f'(W_1 W_2) + W_2 W_1 f''(W_1 W_2),$$

or

$$0 = f'(W) + Wf''(W).$$

Hence $$\log f'(W) + \log W = \text{constant},$$

$$\therefore\ f'(W) = \frac{k}{W},$$

$$\therefore\ f(W) = k \log W + C.$$

Substituting in (1),

$$k \log W_1 + C + k \log W_2 + C = k \log (W_1 W_2) + C,$$

$$\therefore\ C = 0,$$

and

$$\phi = k \log W.$$

This general result was first found for a perfect gas by Boltzmann[1].

Since $dE = td\phi$, where t is the temperature,

$$dE = t\,d\,(k \log W),$$

$$\therefore\ \frac{1}{kt} = \frac{d}{dE}(\log W).$$

The right-hand side can be calculated[2] for a perfect gas and is found to be equal to $\frac{1}{Rt}$, where R is the gas constant per molecule $= 13{\cdot}8 \times 10^{-17}$, so that $k = R$, and $\frac{1}{Rt} = \frac{d}{dE}(\log W)$, in general.

8. *Planck's oscillator.* It has been mentioned above that the mechanism of emission and absorption of radiation by any substance is on the electron theory supposed to be a very large number of 'oscillators' each having its own period and contained in the radiating body. On account of Kirchhoff's laws that the nature of black-body radiation is independent of the substance which emits or absorbs it, the nature of the oscillator is immaterial; Planck chose the simplest form of oscillator, namely, a linear one whose kinetic and potential

[1] L. Boltzmann, 'Vorlesungen über Gastheorie,' 1, § 6.

[2] J. H. Jeans, 'The dynamical theory of gases,' chap. v.

energies are $\frac{1}{2}a\dot{q}^2$ and $\frac{1}{2}bq^2$. The equation of motion is $a\ddot{q} + bq = 0$, so that the period is $2\pi\sqrt{\frac{a}{b}} = \frac{1}{\nu}$, where ν is the frequency of oscillation; therefore $b = 4\pi^2 a\nu^2$,

$$\therefore\ \ddot{q} + 4\pi^2\nu^2 q = 0,$$

$$\therefore\ q = A\cos(2\pi\nu t + \alpha), \quad \ldots\ldots\ldots\ldots\ldots(1),$$

where A, α are constants.

The energy ϵ of the oscillator

$$= \tfrac{1}{2}(a\dot{q}^2 + bq^2)$$

$$= \tfrac{1}{2}a(\dot{q}^2 + 4\pi^2\nu^2 q^2)$$

$$= 2\pi^2 a A^2 \nu^2, \text{ using (1).}$$

Let p be the momentum corresponding to q, so that $p = \frac{\partial T}{\partial \dot{q}}$, where T is the kinetic energy; or, $p = a\dot{q}$.

Then $$\epsilon = \tfrac{1}{2}(a\dot{q}^2 + bq^2) = \frac{p^2}{2a} + 2a\pi^2\nu^2 q^2 \quad \ldots\ldots\ldots(2).$$

The 'phase space' of Gibbs[1] is obtained by using the rectangular coordinates p, q and (2) shows that the 'phase path' of energy ϵ is an ellipse whose semiaxes are

$\sqrt{2a\epsilon}$, $\sqrt{\frac{\epsilon}{2a\pi^2\nu^2}}$ and whose area therefore is $\pi\sqrt{\frac{2a\epsilon^2}{2a\pi^2\nu^2}} = \frac{\epsilon}{\nu}$.

In the classical theory of dynamics all values of ϵ are possible so that these ellipses fill the whole of the phase space. Planck's new idea, which is the root of the quantum theory, was to limit the number of possible phase paths by the condition that the areas of the ellipses must be whole multiples of a constant h.

Hence the area of an ellipse $= nh$, where n is an integer,

or $$\frac{\epsilon}{\nu} = nh,$$

or $$\epsilon = n(h\nu).$$

Thus an oscillator can only contain an integral number of 'quanta' of energy, each quantum being equal to $h\nu$, so that the 'oscillators' in the radiating body can only emit or absorb

[1] J. W. Gibbs, 'Elementary Principles in Statistical Mechanics,' 1902.

energy in 'quanta' of amount $h\nu$. This is the fundamental principle of the quantum theory.

It may be noticed that the above 'quantum condition' that the area of a phase ellipse must $= nh$ can be expressed in the form $\int p\,dq = nh$, where the integral is taken through a complete oscillation of q.

Further since $\epsilon = 2\pi^2 a\nu^2 A^2$,

and $\epsilon = nh\nu$,

$$\therefore\ A = \sqrt{\frac{nh}{2\pi^2 a\nu}},$$

so that a discrete set of possible amplitudes of oscillation is disclosed, corresponding to $n = 1, 2, 3, \ldots$.

9. *The proof of Planck's formula*[1]. Suppose the body contains N_1 oscillators each of frequency ν_1, N_2 of frequency ν_2, and so on, where in the limit $\nu_1, \nu_2, \ldots$ form a continuous sequence.

Consider a typical set containing N oscillators of frequency ν.

If E is the total energy of the oscillators, $\frac{E}{h\nu}$ is the number P of elements of energy to be distributed amongst the N oscillators.

$$\therefore\ E = h\nu P \quad \ldots\ldots\ldots\ldots\ldots\ldots\ldots(1).$$

The number of ways in which this can be done will be the number of 'complexions' corresponding to the state E, i.e. will be the probability W of that state. But this number is

$$\frac{(P+N)!}{N!\,P!},$$

$$\therefore\ W = (P+N)!\big/N!\,P!,$$

$$\therefore\ \log W = \log\{(P+N)!\} - \log(N!) - \log(P!).$$

By Stirling's theorem for very large numbers,

$$n! = \sqrt{2\pi n}\left(\frac{n}{e}\right)^n,$$

[1] M. Planck, Verhandl. d. Deutschen Phys. Ges. 2, p. 237, 1900.

where e is the base of logarithms,

$$\therefore \quad \log (n!) = \tfrac{1}{2} \log (2\pi n) + n \log n - n,$$

and the first term is negligible compared with the others, so that

$$\log (n!) = n \log n - n.$$

Here N, P are large, so that

$$\log W = (P+N) \log (P+N) - (P+N) - N \log N + N - P \log P + P,$$

or

$$\log W = (P+N) \log (P+N) - N \log N - P \log P.$$

But by § 7, $\dfrac{1}{Rt} = \dfrac{d}{dE} (\log W) = \dfrac{1}{h\nu} \dfrac{d}{dP} (\log W)$, using (1),

$$\therefore \quad \frac{h\nu}{Rt} = \log (P+N) + 1 - \log P - 1$$

$$= \log \left(\frac{P+N}{P}\right),$$

$$\therefore \quad P + N = Pe^{h\nu/Rt},$$

$$\therefore \quad P = \frac{N}{e^{h\nu/Rt} - 1},$$

$$\therefore \quad E = h\nu P = \frac{Nh\nu}{e^{h\nu/Rt} - 1}.$$

Thus the mean energy of an oscillator is $\dfrac{h\nu}{e^{h\nu/Rt} - 1}$, or $Rt\left(\dfrac{x}{e^x - 1}\right)$, where $x = \dfrac{h\nu}{Rt}$.

This is the total energy Rt (kinetic and potential) of a molecule on the classical theory, multiplied by the factor

$$\frac{x}{e^x - 1}.$$

For small values of x this factor $\to 1$, so that the quantum and classical theory agree for small values of x, i.e. for large values of λt $\left(\text{since } x = \dfrac{h\nu}{Rt} = \dfrac{hc}{\lambda Rt}, \text{ where } c \text{ is the velocity of radiation}\right)$.

Thus it is for short wave lengths and low temperatures that the two theories diverge.

In 1903 Lorentz[1] calculated the emissive and absorptive powers of a thin metallic plate for long wave lengths and hence found that $E_\lambda = \frac{8\pi Rt}{\lambda^4}$, where R is the molecule gas constant.

Now if $N_\lambda d\lambda$ is the number of oscillators whose wave lengths lie between λ and $\lambda + d\lambda$, the energy $E_\lambda d\lambda$ associated with them is, on Planck's theory, given by $E_\lambda = N_\lambda Rt\left(\frac{x}{e^x - 1}\right)$, where $x = h\nu/Rt$, and this should agree with Lorentz's result for large values of λ, i.e. small values of x,

$$\therefore \quad N_\lambda \frac{Rtx}{e^x - 1} \text{ should} \rightarrow \frac{8\pi Rt}{\lambda^4} \text{ as } x \rightarrow 0,$$

$$\therefore \quad N_\lambda = \frac{8\pi}{\lambda^4}.$$

Thus the number of 'oscillators' emitting waves of lengths between λ and $\lambda + d\lambda$ is $8\pi\lambda^{-4}.d\lambda$.

Finally, $$E_\lambda = \frac{N_\lambda h\nu}{e^{h\nu/Rt} - 1} = \frac{8\pi h\nu}{\lambda^4 (e^{h\nu/Rt} - 1)}.$$

This is Planck's formula for the distribution of energy in the black-body spectrum.

10. *The maximum of E_λ.*

$$E_\lambda = \frac{8\pi h\nu}{\lambda^4 (e^{h\nu/Rt} - 1)} = \frac{8\pi hc}{\lambda^5 (e^{hc/\lambda Rt} - 1)}.$$

If $y = E_\lambda$ and $x = \lambda$, this formula is $y = \frac{8\pi hc}{x^5 (e^{a/x} - 1)}$, where $a = \frac{hc}{Rt}$.

The maximum of y for a given temperature is given by the minimum of $x^5 (e^{a/x} - 1)$ or by

$$5x^4 (e^{a/x} - 1) - x^5 e^{a/x} (a/x^2) = 0,$$

[1] H. A. Lorentz, Proc. Akad. van Wetenschappen Amsterdam, p. 666, 1903; 'Theory of Electrons,' pp. 80–90.

or $$1 - e^{-\theta} = \tfrac{1}{5}\theta,$$

where $$\theta = \frac{a}{x} = \frac{hc}{\lambda R t} = \frac{h\nu}{Rt}.$$

The root of this equation is approximately 4·9651, so that $\frac{hc}{\lambda_m R t} = 4{\cdot}9651$, where λ_m is the value of λ corresponding to the maximum of E_λ.

$$\therefore\ \lambda_m t = \frac{hc}{R\,(4{\cdot}9651)}.$$

But Lummer and Pringsheim found by experiment that

$$\lambda_m t = {\cdot}294 \text{ cm.-deg.},$$

so that $$h = \frac{R\,(4{\cdot}9651)\,({\cdot}294)}{c},$$

where R, the molecule gas constant, is $13{\cdot}8 \times 10^{-17}$.

$$\therefore\ h = \frac{(13{\cdot}8 \times 10^{-17})\,(4{\cdot}9651)\,({\cdot}294)}{3 \times 10^{10}}$$
$$= 6{\cdot}71 \times 10^{-27}.$$

This constant h has been found in other very diverse ways from observations of spectra, the photo-electric effect, and the specific heats of solid bodies. The most probable value of the constant is $6{\cdot}55 \times 10^{-27}$ ergs-sec.[1]

The quantum for the yellow light of the D line of sodium ($\lambda = 5890$ Å.) is

$$h\nu = \frac{hc}{\lambda} = \frac{6{\cdot}55 \times 10^{-27} \times 3 \times 10^{10}}{5890 \times 10^{-8}} = 3{\cdot}3 \times 10^{-12} \text{ ergs.}$$

For a hard X-ray, such as that corresponding to the K_a line for zinc, $\lambda = 1430X = 1430 \times 10^{-11}$ cm. The quantum $h\nu$ for this ray is $1{\cdot}4 \times 10^{-8}$ ergs, or 4000 times the quantum for the yellow light of the sodium D line.

[1] R. A. Millikan, Phys. Review, **7**, p. 355, 1916; Phil. Mag. **34**, p. 16, 1917.

CHAPTER II

THE NUCLEAR THEORY OF THE ATOM; BOHR'S THEORY OF THE HYDROGEN SPECTRUM

11. *The nuclear theory of the atom.* The modern theory of the structure of the atom is in the first place due to J. J. Thomson. His researches on the conduction of electricity through gases[1] in 1897 disclosed the existence of negatively electrified particles, now called 'electrons,' whose mass is 1/1845 of the mass of an atom of hydrogen and whose charge is $4{\cdot}774 \times 10^{-10}$ E.S.U.; the mass and charge of an electron were independent of the nature of the gas used.

It thus became apparent that electrons might be an ultimate constituent of all atoms; Thomson[2] at first proposed a model of the atom in which electrons were in equilibrium in a uniformly diffused positive charge forming a minute sphere, and later[3] gave a kinetic theory in which the electrons were revolving in orbits under the controlling influence of the positive charge. (The discovery of radio-activity had by this time shown that within the atom there was a vast store of kinetic energy.)

At the same time as the electrons, there are produced in the gas tube positively electrified particles, now called 'positive rays' of the same order of mass as the atom of the gas used.

Thomson[4] examined the positive rays from different substances by a deflexion method in which positive particles of a given kind produced a parabola on a photographic plate whatever their speed; the knowledge of these rays has since

[1] J. J. Thomson, 'Conduction of Electricity through Gases,' 1903.
[2] J. J. Thomson, 'Electricity and Matter' (Silliman Lectures), 1903.
[3] J. J. Thomson, 'The Corpuscular Theory of Matter' (Royal Institution Lectures), 1906.
[4] J. J. Thomson, 'Rays of Positive Electricity,' 1913.

been greatly advanced by an experimental method of great power (the 'mass' spectrograph) due to Aston[1].

The researches of Rutherford[2] and his school in which the instrument of the α-particle was used to disclose the nature of the atom, have led to the atomic model proposed by Rutherford which is now generally used in theoretical work.

Rutherford's atom consists of a central positively charged nucleus around which revolve a number of negatively charged electrons. In a neutral atom the total charge of the electrons is equal and opposite to that of the nucleus. The dimensions of the nucleus are small compared with the dimensions of the orbits of the electrons and almost the entire mass of the atom is concentrated in the nucleus. If the number of revolving electrons in the neutral atom is Z, the nuclear charge is Ze, where $-e$ is the charge of an electron.

If the elements are arranged in a table in the order of their atomic weights and numbered successively, starting with hydrogen as 1, the number associated with a given element is its 'atomic number.'

The work of Moseley[3] on X-ray spectra, and later work, such as the direct determination of the nuclear charge for certain metals by Chadwick[4], have shown that if the magnitude e of the electron charge is taken as the unit, the nuclear charge is equal to the atomic number; for atomic number Z, the nuclear charge is Ze.

On this theory the discharge through the gas tube in Thomson's experiments detaches one or more electrons from the atom, leaving the nucleus and the remaining electrons which constitute a positive ray particle.

An atom of hydrogen, whose atomic number is 1, consists of an electron (charge $-e$, mass m) revolving round a nucleus

[1] F. W. Aston, Phil. Mag. **38**, p. 707, 1919.

[2] E. Rutherford, 'Radioactive Substances and their radiations,' 1913, and numerous papers in the Phil. Mag. from 1911 onwards. See also 'A discussion on the structure of the atom,' Proc. Roy. Soc. **90** A, 1914.

[3] H. G. J. Moseley, Phil. Mag. **26**, p. 1024, 1913, and **27**, p. 703, 1914.

[4] J. Chadwick, Phil. Mag. **40**, p. 734, 1920.

(charge $+e$, mass M), where M is about $1845m$, so that the nucleus effectively contains the mass of the atom.

This nucleus, detached from its revolving electron, is a positive ray of hydrogen. This is called a 'proton' and is denoted by H_+.

The nuclei of heavier atoms are built up of protons and electrons. For example, if an element has atomic weight a and atomic number n, the nucleus of its atom will contain a protons to give it the necessary mass, and if the atom is neutral there would be n electrons circulating round it. To make the charge on the nucleus equal and opposite to that of the electrons, the nucleus must also contain $(a-n)$ electrons, which would not affect its mass appreciably. Thus the nucleus would be an aggregate of a protons and $(a-n)$ electrons of total charge ne and there would be revolving electrons of total charge $-ne$.

The spectrum and chemical properties of the element depend upon the arrangement of its outer electrons. Thus two atoms, say one with a nucleus (22 protons, 12 electrons) and 10 revolving electrons, and the other with a nucleus (20 protons, 10 electrons) and 10 revolving electrons, would have identical spectroscopic (but see § 123) and chemical properties. But the atomic weight of the former would be 22 and of the latter 20. Such atoms are called 'isotopes.' The atomic weight of the gas neon is 20·2, and Aston[1] has shown that the gas is a mixture of two isotopes of atomic weights 20, 22.

Isotopes are not separable by chemical methods; they can be separated mechanically, as in the mass spectrograph, by what is effectively a centrifugal method depending on their difference of mass.

The next element after hydrogen is helium, atomic number 2, atomic weight 4. Thus its nucleus is (4 protons, 2 electrons), and in the neutral atom (He) there are two revolving electrons. If it is singly 'ionised' so as to lose one electron, it becomes He_+ and the atom has a total charge $+e$; if doubly ionised so as to lose its two electrons, it becomes He_{++} and the atom

[1] F. W. Aston, 'Isotopes,' 1922.

has a total charge 2*e*. This helium nucleus, He_{++}, is the α-particle emitted from radioactive substances. These can be represented diagrammatically as follows, though the figure is only intended to indicate the existence of orbits but not their nature.

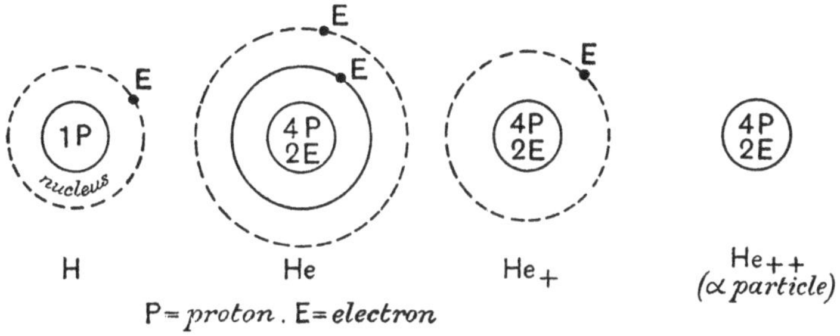

Lithium, the next element, has atomic weight 6·94, atomic number 3.

Aston has shown this to consist of two isotopes of atomic weights 6, 7 (Li_6, Li_7).

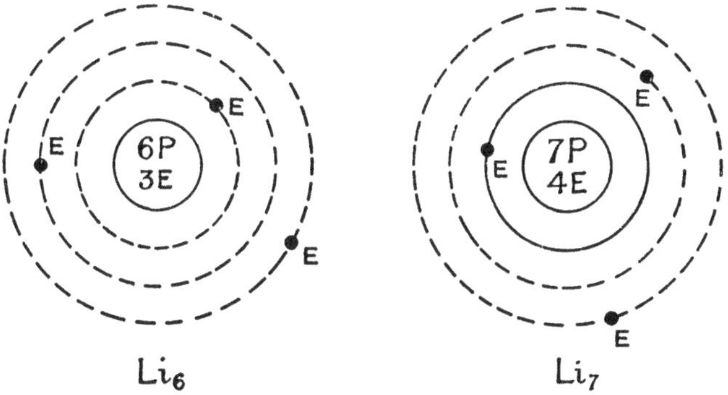

12. *The line spectrum of hydrogen.* The visible line spectrum of hydrogen contains a series of lines at intervals which decrease rapidly as λ decreases; they crowd together in the violet region and tend to a limit—'the head of the series.' This series of lines is known as Balmer's series. In

1885, Balmer[1] noticed that the wave lengths of the first four lines H_α, H_β, H_γ, H_δ were in the ratio

$$\tfrac{9}{5} : \tfrac{4}{3} : \tfrac{25}{21} : \tfrac{9}{8} \text{ nearly,}$$

or

$$\tfrac{9}{5} : \tfrac{16}{12} : \tfrac{25}{21} : \tfrac{36}{32},$$

or

$$\frac{3^2}{3^2-2^2} : \frac{4^2}{4^2-2^2} : \frac{5^2}{5^2-2^2} : \frac{6^2}{6^2-2^2}.$$

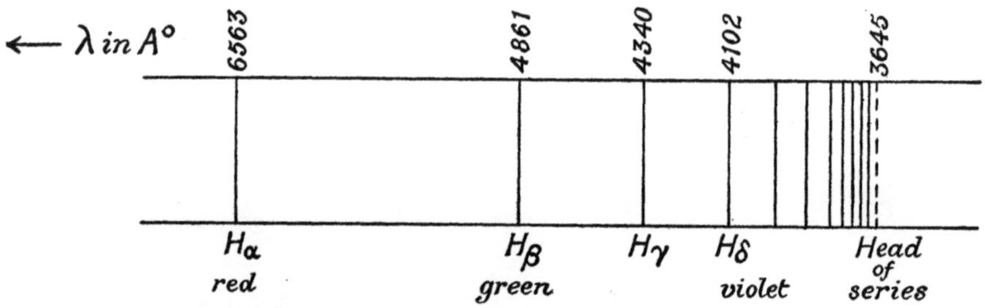

This suggested the formula

$$\lambda \propto \frac{n^2}{n^2-2^2}, \qquad n = 3, 4, 5, \ldots,$$

or

$$\frac{1}{\lambda} \propto 1 - \frac{2^2}{n^2},$$

or

$$\frac{1}{\lambda} = R\left(\frac{1}{2^2} - \frac{1}{n^2}\right),$$

where R is a constant. Using the known values of λ in cm. for these lines, the formula is found to be very exact indeed, with $R = 109678$.

The 'wave number' N of a line in spectroscopy is the reciprocal of the wave length λ, or $N = \dfrac{1}{\lambda}$,

so that

$$N = R\left(\frac{1}{2^2} - \frac{1}{n^2}\right).$$

If ν is the 'frequency' of the line, $\left(\dfrac{1}{\nu}\right) c = \lambda$, where c is 3×10^{10} cm. per sec., the velocity of light,

$$\therefore\ \nu = \frac{c}{\lambda} = cN.$$

[1] J. J. Balmer, Ann. d. Phys. 25, p. 80, 1885.

Balmer's formula is then also $\nu = cR\left(\frac{1}{2^2} - \frac{1}{n^2}\right)$.

Balmer conjectured that the formula $\frac{1}{\lambda} = R\left(\frac{1}{2^2} - \frac{1}{n^2}\right)$ might be a particular case of $\frac{1}{\lambda} = R\left(\frac{1}{s^2} - \frac{1}{n^2}\right)$, where s in any integer less than n, and that there might be undiscovered lines satisfying this formula.

This simple formula had naturally resulted in many attempts at a theoretical explanation, but without success until, in 1913, Niels Bohr[1], by a new concept based upon Planck's quantum theory, gave a theory of the hydrogen spectrum which not only led to the Balmer formula but also to an expression for the Rydberg constant R in terms of known physical constants, which was in striking numerical agreement with their known values.

13. *Classical theory.* Rutherford's H atom is an electron $(m, -e)$ revolving round a nucleus $(M, +e)$, where $M = 1845m$. The ratio of the electric force to the gravitational force on the electron is $e^2/\gamma Mm$, where γ is the constant of gravitation equal to 666×10^{-10}. This ratio is

$$(4{\cdot}77 \times 10^{-10})^2/666 \times 10^{-10}\,(1845)\,(9{\cdot}04 \times 10^{-28})^2 = 2 \times 10^{39},$$

so that the gravitational force may be neglected.

Suppose the orbit of the electron to be a circle of radius a with its centre at the nucleus and neglect the motion of the massive nucleus.

Let ω be the frequency, i.e. the number of revolutions per second, so that the velocity v is $2\pi a\omega$.

The equation of motion is

$$\left.\begin{aligned} m\left(\frac{v^2}{a}\right) &= \frac{e^2}{a^2} \\ \text{or}\qquad 4\pi^2 m a^3 \omega^2 &= e^2 \end{aligned}\right\} \quad \ldots\ldots\ldots\ldots\ldots\ldots\ldots(1).$$

Let W be the work required to remove the electron from its orbit to a position of rest outside the atom (effectively

[1] N. Bohr, Phil. Mag. **26**, pp. 1, 476, 857, 1913.

at infinity for this scale of magnitudes), i.e. the work of 'ionisation.' Then if the energy of the system with the electron at rest outside the atom is taken to be the zero, $(-W)$ is the energy in the orbit, and

$$-W = \tfrac{1}{2}mv^2 - \frac{e^2}{a},$$

$$\therefore\ W = \frac{e^2}{a} - \tfrac{1}{2}mv^2 = \frac{e^2}{a} - \frac{1}{2}\frac{e^2}{a}, \text{ from (1)},$$

$$= \frac{e^2}{2a},$$

$$\therefore\ a = \frac{e^2}{2W} \text{ and } \omega^2 = \frac{e^2}{4\pi^2 m}\left(\frac{2W}{e^2}\right)^3 = \frac{2W^3}{\pi^2 m e^4} \quad \ldots(2).$$

The theory of electrodynamics requires an accelerated electron, such as this is, to radiate energy, so that in obtaining these equations it has been assumed that the effect on the motion due to this emission of radiation is at any moment small compared with the effect of the electric force between the electron and nucleus. But after many revolutions in the orbit (the period is only of order 10^{-8} sec. in the normal H atom) this small loss of energy due to radiation would tell, and W (the negative energy) would be perceptibly greater; so that a would diminish and ω increase, as indicated by the formula (2). Assuming that the frequency ω in the orbit is also the frequency of the corresponding spectral line, there would be hydrogen atoms in varying states having all possible frequencies and the result would be a continuous spectrum, not a line spectrum.

14. *Bohr's theory of the hydrogen spectrum.* Bohr introduced the assumption that in the atom only certain orbits are possible which have a peculiar kind of stability, such that it is impossible for the atom to emit or receive energy except by a process involving a complete transition from one to another of these orbits. These states of the atom he called 'stationary states,' and he supposed that the atom can exist in any of these states without emitting radiation. These states were to be determined by the condition that only such orbits were permissible as had angular momentum which was an integral multiple of $\frac{h}{2\pi}$, where h is Planck's constant, this being a natural development for orbits of Planck's quantum condition for the oscillator, which had also been proposed by Nicholson[1] a year before. But Bohr's new concept[2] which was to be the key to the solution of the problem of spectra, was that the radiation emitted during a transition between two stationary states has a frequency ν given by the relation $E - E' = h\nu$, where h is Planck's constant and E, E' are the values of the energy in the states before and after the transition. This frequency is quite different from the frequency of revolution in either of the two states.

From the Nicholson-Bohr condition that the angular momentum is $nh/2\pi$, where n is 1, 2, 3, ..., it follows that

$$mav = nh/2\pi,$$

or

$$4\pi^2 ma^2\omega = nh \quad \text{.....................(3).}$$

Using $4\pi^2 ma^3\omega^2 = e^2$, already found as equation (1), a and ω are determined.

The values are

$$a = \frac{h^2n^2}{4\pi^2me^2}, \quad \omega^2 = \frac{4\pi^2me^4}{h^3n^3}.$$

Thus the possible stationary states are orbits of radii proportional to $1^2, 2^2, 3^2, \ldots$, and n is the 'quantum number' of the orbit.

Also

$$W = \frac{e^2}{2a} = \frac{2\pi^2me^4}{h^2n^2}.$$

Using Bohr's condition $E - E' = h\nu$, we have $W' - W = h\nu$, since W is the negative energy and is equal to $-E$, so that

$$h\nu = \frac{2\pi^2me^4}{h^2}\left(\frac{1}{n'^2} - \frac{1}{n^2}\right),$$

where n, n' are the quantum numbers of the initial and final orbits.

[1] J. W. Nicholson, Monthly Notices R.A.S. **72**, p. 677, 1912.

[2] See also N. Bohr, 'The Theory of Spectra and Atomic Constitution' (Essay I), published by the Cambridge University Press, 1922.

For an emission of energy, n changes to a smaller number n', so that the new orbit has a smaller radius than the old one.

From the above, $\nu = \frac{2\pi^2 me^4}{h^3}\left(\frac{1}{n'^2} - \frac{1}{n^2}\right)$.

This is the generalised Balmer formula

$$\frac{1}{\lambda} = R\left(\frac{1}{n'^2} - \frac{1}{n^2}\right),$$

or

$$\nu = Rc\left(\frac{1}{n'^2} - \frac{1}{n^2}\right),$$

provided

$$R = \frac{2\pi^2 me^4}{ch^3}.$$

Using the known values

$$m = 9{\cdot}04 \times 10^{-28} \text{ gr.},$$
$$c = 3 \times 10^{10} \text{ cm. per sec.},$$
$$e = 4{\cdot}77 \times 10^{-10} \text{ E.S.U.},$$
$$h = 6{\cdot}55 \times 10^{-27} \text{ ergs. sec.},$$

the theoretical value of R was found to be 109,700, the experimental constant being 109,677. This amazing verification at once fixed attention upon the quantum theory, which up to then had received sceptical regard from physicists in general[1].

15. *Ritz's combination principle.* In 1908, Ritz[2] had put forward the principle that 'the wave number of any spectral line is the difference of those of two other spectral lines for a given substance.' Thus for H_α, H_β, H_γ, Balmer's formula gave the wave number N as

$$R\left(\frac{1}{2^2} - \frac{1}{3^2}\right),\quad R\left(\frac{1}{2^2} - \frac{1}{4^2}\right),\quad R\left(\frac{1}{2^2} - \frac{1}{5^2}\right).$$

The difference between the wave numbers of H_β and H_α is $R\left(\frac{1}{3^2} - \frac{1}{4^2}\right)$, and between those of H_γ and H_α is $R\left(\frac{1}{3^2} - \frac{1}{5^2}\right)$. Ritz suspected the existence of these two lines which on

[1] British Assoc. Report, 'Discussion on the Quantum Theory' by Section A, 1913.

[2] W. Ritz, Phys. Zeitschr. **9**, p. 521, 1908.

account of their small wave number must be in the infra-red; they were looked for and observed by Paschen[1] in 1908. These are the first two lines of the generalised Balmer formula $N = R\left(\frac{1}{n'^2} - \frac{1}{n^2}\right)$, where $n' = 3$, $n = 4, 5, 6, \ldots$, which gives this Paschen-Ritz series of lines.

After the publication of Bohr's theory, the lines

$$N = R\left(\frac{1}{1^2} - \frac{1}{n^2}\right), \quad n = 2, 3, 4, \ldots,$$

indicated by his theory, but not deducible from the Balmer lines by Ritz's principle, were looked for and discovered by Lyman[2], who found the first two of these lines in the ultra-violet. This series of lines is known as the Lyman series.

Since then three more of the Paschen-Ritz lines have been found by Brackett[3], who also found lines of the series $N = R\left(\frac{1}{4^2} - \frac{1}{n^2}\right)$ for $n = 5, 6$, all of these in the infra-red.

16. *Energy levels in the atom.* The negative energy W in the orbit of quantum number n was found to be $\frac{2\pi^2 me^4}{h^2 n^2}$ or $\frac{Rch}{n^2}$.

Thus each orbit represents an energy value or 'energy level' in the atom. The negative energy W corresponding to a given orbit or level is the work to be done on an electron to drive it from that orbit to the periphery of the atom at which $W = 0$. If V volts is the E.M.F. which would just do this work on the electron, then since

$$V \text{ volts} = \frac{V}{300} \text{ E.S.U.},$$

$$\left(\frac{V}{300}\right) e = W, \quad \text{or} \quad V = \frac{300W}{e}.$$

Thus to each W there is a corresponding voltage V, so that an energy level is usually measured in volts, the term having the meaning indicated above.

1 F. Paschen, Ann. d. Phys. **27**, p. 565, 1908.
2 T. Lyman, Nature, **93**, p. 241, 1914.
3 F. S. Brackett, Nature, **109**, p. 209, 1922.

If N is the wave number of the line emitted by an electron falling from the edge of the atom to the orbit whose quantum number is n and negative energy W,

$$N = R\left(\frac{1}{n^2} - \frac{1}{\infty}\right) = \frac{R}{n^2},$$

and since $$W = \frac{Rch}{n^2}, \quad N = \frac{W}{hc} = \frac{eV}{300hc}.$$

This is the wave number corresponding to the energy level V volts. Inserting the values of e, h, c given above (p. 24), we have $N = 8100V$.

The wave number of a spectral line emitted by a transition from an orbit n_1 to an orbit n_2 is

$$= R\left(\frac{1}{{n_2}^2} - \frac{1}{{n_1}^2}\right) = N_2 - N_1,$$

and is thus the difference of the aforesaid wave numbers assigned to the two orbits in question.

17. *The normal state of the hydrogen atom.* The orbit of least energy, i.e. of greatest W, represents the normal state of the atom; this orbit, called the 'ground orbit,' corresponds therefore to $n = 1$ and its radius a is $\frac{h^2}{4\pi^2 me^2}$. Putting in the known values of h, m, e, $a = 5{\cdot}28 \times 10^{-9}$ cm., and is of the same order as the radii of atoms calculated from the kinetic theory of gases. If the atom is excited by some influence, such as a flame or the electric arc, the electron is driven from the ground orbit to an orbit of higher quantum number, and by spontaneous return to orbits of lower number causes the emission of spectral lines. The proportion of the transitions of a given kind in the whole series of atoms under the exciting influence determines the intensity of the spectral line corresponding to that transition.

18. *Observation of the Balmer lines of high order.* The radius of the nth orbit is $\frac{h^2 n^2}{4\pi^2 me^2} = n^2\,(5{\cdot}28 \times 10^{-9})$ cm.

If n is large, the radius is large and it may not be possible to reduce the pressure on the gas in a laboratory sufficiently for such large atoms to exist, whereas in the stars or nebulae pressures low enough may exist.

Thus while in the laboratory the 20th line of the Balmer series was the highest observed by Wood[1], in the sun's chromosphere the 29th line had been observed by Dyson[2]. Dyson's extreme line corresponds to a transition from $n = 31$ to $n = 2$, so that the radius of the outer orbit was $(31)^2 (5{\cdot}28 \times 10^{-9})$ or 5×10^{-6}, which is comparable with the space between the atoms of a gas in its normal state. The theory too is borne out by Wood's observation that if the pressure of the gas was increased the Balmer lines of higher order disappeared one by one.

19. *The spectrum of ionised helium* (He_+). The atom of ionised helium is a neutral He atom which has lost one electron. The nucleus has a charge $2e$ and a single electron revolves round it. For a nucleus of charge Ze with one electron circulating round it (i.e. a neutral atom of atomic number Z which has lost all its electrons save one), it is evident that the results just found for the hydrogen atom still hold if e^2 is replaced by $(Ze)\,e$ or Ze^2.

The spectrum is given by

$$\frac{1}{\lambda} = \frac{2\pi^2 m\,(Ze^2)^2}{ch^3}\left[\frac{1}{n'^2} - \frac{1}{n^2}\right],$$

or

$$\frac{1}{\lambda} = RZ^2\left(\frac{1}{n'^2} - \frac{1}{n^2}\right).$$

In the case of He_+, $Z = 2$ and

$$\frac{1}{\lambda} = 4R\left(\frac{1}{n'^2} - \frac{1}{n^2}\right).$$

The series $n' = 1$, $n = 2, 3, 4, \ldots$ lies far in the ultra-violet and has not so far been observed.

Of the series $n' = 2$, $n = 3, 4, \ldots$, the first two lines have been observed by Lyman[3], also in the ultra-violet.

[1] R. W. Wood, Proc. Roy. Soc. **97** A, p. 455, 1920.
[2] F. W. Dyson, Proc. Roy. Soc. **68** A, p. 33, 1901.
[3] T. Lyman, Nature, **104**, p. 314, 1919.

The series $n' = 3$, $n = 4, 5, 6, \ldots$ are

$$\frac{1}{\lambda} = 4R\left(\frac{1}{3^2} - \frac{1}{4^2}\right), \quad 4R\left(\frac{1}{3^2} - \frac{1}{5^2}\right), \ldots,$$

$$\text{or } R\left\{\frac{1}{(1\tfrac{1}{2})^2} - \frac{1}{2^2}\right\}, \quad R\left\{\frac{1}{(1\tfrac{1}{2})^2} - \frac{1}{(2\tfrac{1}{2})^2}\right\}, \quad R\left\{\frac{1}{(1\tfrac{1}{2})^2} - \frac{1}{3^2}\right\}, \ldots,$$

$$\text{or} \qquad R\left\{\frac{1}{(1\tfrac{1}{2})^2} - \frac{1}{n^2}\right\},$$

where $n = 2, 2\tfrac{1}{2}, 3, 3\tfrac{1}{2}, \ldots$.

These were discovered by Fowler[1] in 1912 in a mixture of H and He, and were attributed to hydrogen by a kind of generalisation of Balmer's formula to which half-integers were admitted, but after the appearance of Bohr's theory they were assigned to He_+. Later they were observed by Evans[2] (1915) and by Paschen[3] (1916) in tubes of He from which all traces of H had been removed so that none of the usual H lines appeared, which confirmed the new theory.

The series $n' = 4$, $n = 5, 6, \ldots$ are known as the Pickering lines.

The terms are

$$\frac{1}{\lambda} = 4R\left(\frac{1}{4^2} - \frac{1}{5^2}\right), \quad 4R\left(\frac{1}{4^2} - \frac{1}{6^2}\right), \quad 4R\left(\frac{1}{4^2} - \frac{1}{7^2}\right), \ldots,$$

$$\text{or} \quad \frac{1}{\lambda} = R\left(\frac{1}{2^2} - \frac{1}{(2\tfrac{1}{2})^2}\right), \quad R\left(\frac{1}{2^2} - \frac{1}{3^2}\right), \quad R\left(\frac{1}{2^2} - \frac{1}{(3\tfrac{1}{2})^2}\right), \ldots.$$

These lines were observed by Pickering[4] in 1896 in the spectrum of ζ Puppis: the even members of this series are the Balmer lines of hydrogen; the odd ones were the new ones noticed by Pickering and assigned to H also by the half-integer modification of Balmer's formula mentioned above.

After Bohr's theory, they were assigned to He_+. Thus He_+ can emit the Balmer lines of H. The fact that in the spectrum of ζ Puppis the Balmer lines are brighter than the new lines Pickering found indicates that the source of the lines is a mixture of hydrogen and helium.

[1] A. Fowler, Monthly Notices R.A.S. **73**, p. 62, 1912.
[2] E. J. Evans, Phil. Mag. **29**, p. 284, 1915.
[3] F. Paschen, Ann. d. Phys. **50**, p. 901, 1916.
[4] E. C. Pickering, Astrophys. Jour. **4**, p. 369, 1896.

20. *The spectrum of doubly ionised lithium* (Li_{++}). The neutral atom has lost two electrons, so that the nucleus (charge $3e$) has one electron revolving round it. Hence

$$\frac{1}{\lambda} = 9R\left(\frac{1}{n'^2} - \frac{1}{n^2}\right).$$

The lines $n' = 6$, $n = 10, 13, 14$ of this series were found by Nicholson[1] in the spectra of the Wolf Rayet stars of O type. In the set $n' = 6$, the Balmer lines of H are included, corresponding to $n = 9, 12, 15, \ldots$.

21. *The effect of the motion of the nucleus.* In Bohr's original theory, the nucleus was supposed to be so massive that its small motion was negligible. To allow for this motion, consider two particles of masses m_1, m_2 moving under a common attraction $f(r)$ along the line joining them, r being their distance apart. Let G be their centre of gravity which has zero acceleration.

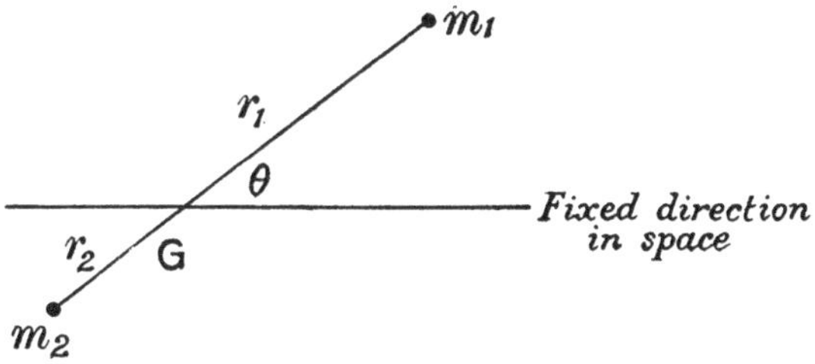

The equations of motion of m_1 are

$$\left.\begin{aligned} m_1(\ddot{r}_1 - r_1\dot{\theta}^2) &= -f(r) \\ r_1^2\dot{\theta} &= k \end{aligned}\right\},$$

where k is a constant.

But $$\frac{r_1}{m_2} = \frac{r_2}{m_1} = \frac{r}{m_1 + m_2} \quad \ldots\ldots\ldots\ldots\ldots\ldots(1),$$

$$\therefore \quad \left.\begin{aligned} \frac{m_1 m_2}{m_1 + m_2}(\ddot{r} - r\dot{\theta}^2) &= -f(r) \\ r^2\dot{\theta} &= k' \end{aligned}\right\},$$

where k' is another constant, or

[1] J. W. NICHOLSON, Monthly Notices R.A.S. **73**, p. 382, 1913.

$$\left.\begin{aligned}\mu(\ddot{r} - r\dot{\theta}^2) &= -f(r)\\ r^2\dot{\theta} &= k'\end{aligned}\right\},$$

where $$\frac{1}{\mu} = \frac{1}{m_1} + \frac{1}{m_2}.$$

Thus the motion of m_1 relative to m_2 is the same as if m_2 were fixed and m_1 were replaced by μ.

The kinetic energy is

$$\tfrac{1}{2}m_1(\dot{r}_1^{\,2} + r_1^{\,2}\dot{\theta}^2) + \tfrac{1}{2}m_2(\dot{r}_2^{\,2} + r_2^{\,2}\dot{\theta}^2) = \frac{1}{2}\frac{m_1 m_2}{m_1 + m_2}(\dot{r}^2 + r^2\dot{\theta}^2),$$

using (1), $$= \tfrac{1}{2}\mu(\dot{r}^2 + r^2\dot{\theta}^2)$$

and is the same as if m_2 were fixed and m_1 were replaced by μ.

Also the angular momentum of the system about G is

$$\tfrac{1}{2}(m_1 r_1^{\,2}\dot{\theta} + m_2 r_2^{\,2}\dot{\theta}) = \tfrac{1}{2}\mu r^2\dot{\theta}$$

and again is the same as if m_2 were fixed and m_1 were replaced by μ.

Hence all the spectral formulae of Bohr can be corrected to allow for the motion of the nucleus if the mass m of the electron is replaced by $\dfrac{mM}{m+M}$, where M is the mass of the nucleus.

Thus for an ionised element of atomic number Z, with one electron only left, the spectrum, allowing for the motion of the nucleus, is given by

$$\frac{1}{\lambda} = \frac{2\pi^2\mu Z^2 e^4}{ch^3}\left(\frac{1}{n'^2} - \frac{1}{n}\right), \qquad (\S\ 19),$$

where $$\mu = \frac{mM_z}{m + M_z},$$

where M_z is the mass of the nucleus, and m that of an electron. Writing

$$\frac{2\pi^2\mu e^4}{ch^3} = R_z, \text{ then } \frac{1}{\lambda} = Z^2 R_z\left(\frac{1}{n'^2} - \frac{1}{n^2}\right).$$

Thus for hydrogen $$\frac{1}{\lambda} = R_{\mathrm{H}}\left(\frac{1}{n'^2} - \frac{1}{n^2}\right),$$

for He_+, $$\frac{1}{\lambda} = 4R_{\mathrm{He}+}\left(\frac{1}{n'^2} - \frac{1}{n^2}\right),$$

for Li_{++} $$\frac{1}{\lambda} = 9R_{Li_{++}}\left(\frac{1}{n'^2} - \frac{1}{n^2}\right),$$

and so on.

From the above it follows that

$$\frac{R_{He_+}}{R_H} = \frac{\mu_{He_+}}{\mu_H} = \frac{M_{He_+}}{m + M_{He_+}} \cdot \frac{m + M_H}{M_H};$$

and since $M_{He_+} = 4M_H$,

$$\frac{R_{He_+}}{R_H} = \frac{M_H + m}{M_H + \frac{1}{4}m} \quad \text{..................(2)}.$$

By accurate observations of the Pickering lines for He_+ and the Balmer lines for H, Paschen in 1916 found that

$$R_{He_+} = 109722{\cdot}1,$$
$$R_H = 109677{\cdot}7.$$

Equation (2) then leads to $\frac{M_H}{m} = 1843{\cdot}7$. This is another striking numerical result arising from Bohr's theory, for this ratio of the mass of the hydrogen atom to the mass of the electron has been found in many other ways. The most probable value[2] of the ratio is 1845.

It is apparent from the above work that the even members of the Pickering series for He_+ are given by

$$\frac{1}{\lambda} = R_{He_+}\left(\frac{1}{2^2} - \frac{1}{3^2}\right),$$

while the Balmer lines for H are given by

$$\frac{1}{\lambda} = R_H\left(\frac{1}{2^2} - \frac{1}{3^2}\right).$$

The 'precision' work of Paschen showed the slight difference between the two series.

22. *The continuous spectrum of hydrogen.* From the 'head' of the Balmer series right up to the ultra-violet there is a continuous spectrum of hydrogen which has been observed both in the laboratory and through the telescope. The 'head'

[1] F. Paschen, Ann. d. Phys. **50**, p. 935, 1916.
[2] R. A. Millikan, Phil. Mag. **34**, p. 16, 1917.

of the Balmer series corresponds to a transition from the orbit $n = \infty$ whose energy is zero to the orbit $n = 2$. If a moving electron from outside is annexed by the atom, and falls into the orbit $n = 2$, more energy is emitted than could be by an electron of zero energy falling from $n = \infty$ to $n = 2$, so that a spectral line of higher frequency than the head of the series is produced; and, as such a moving electron may have any velocity and so any energy, the spectrum beyond the 'head' is continuous.

CHAPTER III

THE CORRESPONDENCE PRINCIPLE; THE PHOTO-ELECTRIC EFFECT; EINSTEIN'S DEDUCTION OF PLANCK'S FORMULA

23. *The correspondence principle of Bohr.* For the hydrogen atom it was found that $\omega = \frac{4\pi^2 me^4}{h^3 n^3}$ for the orbit of quantum number n. The Rydberg constant R is equal to $\frac{2\pi^2 me^4}{ch^3}$, so that

$$\omega = \frac{2Rc}{n^3} \quad \text{.........................(1).}$$

The frequency ν of a spectral line due to a transition from n to n' is equal to $Rc\left(\frac{1}{n'^2} - \frac{1}{n^2}\right)$. If n, n' are very large we may write $n' = n - \Delta n$, where Δn is an integer small compared with n or n',

$$\therefore \ \nu = Rc\left[\frac{1}{(n-\Delta n)^2} - \frac{1}{n^2}\right] = Rc\left[\frac{1}{n^2}\left(1 + \frac{2\Delta n}{n}\right) - \frac{1}{n^2}\right] \text{ approx.}$$

$$= \frac{2Rc}{n^3}\,.\,\Delta n = \omega\,.\,\Delta n \text{ from (1).}$$

In the orbital motion the electron has a frequency ω or period $1/\omega$, so that its coordinates can be expressed as a Fourier series of the type $\Sigma C_\tau \cos(2\pi\tau\omega t + \epsilon_\tau)$, where $\tau = 1, 2, 3, \ldots$ and C_τ, ϵ_τ are constants, and on the classical theory the frequencies of these simple harmonic terms, ω, 2ω, 3ω, ..., are the frequencies of the emitted radiation corresponding to this orbit. The formula $\nu = \omega\,.\,\Delta n$, since Δn is an integer, gives precisely these frequences ω, 2ω, 3ω, ... when n is large. Thus for large quantum numbers the frequencies given by the classical and the quantum theory are the same.

But there is this difference, that while the frequencies

$\omega, 2\omega, 3\omega, \ldots$ are all emitted together on the classical theory by one process, namely, the description of a single orbit, according to Bohr's theory the emission of each of $\omega, 2\omega, 3\omega, \ldots$ is an independent process, for on that theory they are due to transitions where $\Delta n = 1, 2, 3, \ldots$, i.e. from orbits of numbers $n+1$, $n+2, \ldots$ to that of number n, where n is large.

On the classical theory, the relative intensities of the lines depend upon the ratios of the coefficients C_1, C_2, $C_3, \ldots$ in the Fourier series; on the quantum theory they must depend upon the relative probabilities of the transitions $n+1 \to n$, $n+2 \to n, \ldots$. Thus when n is large the probability of a particular transition may be taken to be proportional to the amplitude of the corresponding harmonic component in the orbital motion.

Bohr assumes that even when the quantum numbers are not large the probability of a transition between two stationary states n_1 and n_2 $(n_1 \sim n_2 = \Delta n)$ depends upon the amplitudes C_1 and C_2 of harmonic components of frequencies $\omega_1 \Delta n$, $\omega_2 \Delta n$ in the respective orbital motions of those states. This is the principle of 'correspondence.' The relation must in general be very complicated, as is that between the frequency of the radiation and the frequencies in the two orbits.

But it may be presumed that if a harmonic component is absent from *both* of the orbital motions concerned in a transition, the probability is zero and the transition will not occur; the corresponding line will not be present in the spectrum.

It is in this last respect that the principle of correspondence has been of so great value in explaining anomalies in the occurrence of spectral lines, as will be seen in later chapters.

For example, in the case of Planck's oscillator

$$q = C \cos (2\pi\omega t + \epsilon).$$

Here there is only one frequency ω, and since $\nu = \omega \Delta n$, Δn can only be 1 so that transitions only occur between orbits whose quantum numbers differ by unity.

24. *The photo-electric effect.* In 1887 Hertz[1] showed that the incidence of ultra-violet light on an uncharged conductor caused it to become positively charged. This was due to the expulsion of electrons from the atoms of the metal by the incident radiation.

The velocities of emission of the electrons are found to range from zero up to a certain maximum v dependent on the *frequency* but not on the intensity of the incident light or on the temperature[2]. For a given metal v increases with the frequency ν, but there is a critical frequency ν_0 below which no emission occurs, whatever the intensity of the radiation.

This effect had resisted all attempts at explanation by the classical theory.

In 1905, Einstein[3], using the new quantum ideas, readily accounted for it on the new theory. He assumed that the incident radiation consisted of 'light quanta,' each of amount $h\nu$, and that the ejection of an electron from an atom is due to the absorption by the atom of one of these quanta. If W is the energy required to liberate an electron from an atom, $h\nu - W$ is used in giving it kinetic energy, so that if v is the velocity of emission $h\nu - W = \frac{1}{2}mv^2$. It is thus apparent that no electron can be liberated unless $h\nu \geqslant W$, so that the frequency must be at least ν_0 ('threshold frequency'), where $h\nu_0 = W$.

Thus $$h(\nu - \nu_0) = \tfrac{1}{2}mv^2.$$

Electrons liberated at the very surface of a metal would have the maximum v; those liberated inside would lose energy in getting out to the surface and have velocities less than this maximum.

[Einstein's theory of 'light quanta' is not now generally accepted by physicists, but the argument above does not essentially depend upon their existence. All that is necessary

[1] H. Hertz, Wied. Ann. **31**, p. 983, 1887. Also 'Electric Waves,' by H. Hertz, p. 63, 1893.

[2] 'Report on radiation and the quantum theory,' by J. H. Jeans, chap. v.

[3] A. Einstein, Ann. d. Phys. **17**, p. 132, 1905, and **20**, p. 199, 1906.

is to assume that interchanges of energy between radiation and atoms can only occur by quanta.]

25. *The X-ray photo-electric effect.* This effect and its converse are observed for X-rays.

If X-rays of frequency ν impinge on matter, electrons are ejected with maximum velocity v, where $\frac{1}{2}mv^2 = h\nu - W$. Now W is of order 5 to 20 volts, while for X-rays of average frequency, $h\nu$ is of order 10,000 volts, so that effectively

$$\tfrac{1}{2}mv^2 = h\nu.$$

If electrons with velocity v impinge on matter, the converse occurs and X-rays are produced of frequency given by the same equation. This latter converse effect occurs in an X-ray bulb when the electrons impinge on the anti-cathode and produce X-rays. In the bulb the maximum velocity of the electrons is determined by the voltage used and therefore the maximum frequency of the X-rays also.

This accounts for the sudden termination at an upper limit of frequency of the continuous X-ray spectrum emitted from the anti-cathode. [This 'impulse' spectrum, as it is called, is to be distinguished from the X-ray line spectrum 'characteristic' of the material of the anti-cathode, which is emitted at the same time as the former spectrum.]

26. *Einstein's deduction of Planck's radiation formula.* The influence of Bohr's correspondence principle is apparent in the deduction of Planck's formula made by Einstein in 1917[1]. As in § 9, it is assumed that the numbers of atoms in states whose energies are $E_1, E_2, \ldots$ are $N_1, N_2, \ldots$. An atom can pass from a state r of energy E_r to a state s of lower energy E_s by emitting radiation of frequency ν_{rs}, where $E_r - E_s = h\nu_{rs}$; and it can pass from the state s to the state r by absorbing radiation of this frequency.

It is assumed that incident radiation of this frequency can induce transitions from the state r to the state s, and also from the state s to the state r (corresponding to emission and absorption on the classical theory); also that the number of

[1] A. Einstein, Phys. Zeitschr. **18**, p. 122, 1917.

atoms passing per second from the state r to the state s is proportional to the number in the state r (i.e. N_r) and to the energy density $\rho\ (\nu_{rs})$ of the radiation causing the change [the density is expressed as a function ρ of the frequency ν_{rs}].

Thus the number passing per second from the state r to the state s is $a_{rs}N_r\rho\ (\nu_{rs})$, where a_{rs} is an atomic constant.

So the number passing from the state s to the state r is $a_{sr}N_s\rho\ (\nu_{rs})$, where a_{sr} is another atomic constant.

Moreover, there is the possibility of spontaneous changes (as in radioactivity) from states of higher to states of lower energy; the number of atoms for which this occurs per second is $b_{rs}.N_r$.

For equilibrium between the atoms and the incident radiation, the number passing per second in one sense must be equal to that in the opposite sense, so that

$$\rho\ (\nu_{rs})\ \{a_{rs}N_r - a_{sr}N_s\} + b_{rs}N_r = 0.$$

This is a formula to determine $\rho\ (\nu_{rs})$.

In the ordinary theory of gases the probability of an energy state E is proportional to $e^{-E/Rt}$, so that N_r varies as $e^{-E_r/Rt}$ (Boltzmann)[1]. Hence $N_r = g_r e^{-E_r/Rt}$, $N_s = g_s e^{-E_s/Rt}$, where g_r, g_s are constants.

Using these we obtain

$$\rho\ (\nu_{rs})\ \{a_{rs}g_r e^{-E_r/Rt} - a_{sr}g_s e^{-E_s/Rt}\} + b_{rs}g_r e^{-E_r/Rt} = 0,$$

or

$$\rho\ (\nu_{rs})\ \{a_{sr}g_r e^{(E_r - E_s)/Rt} - a_{rs}g_s\} = b_{rs}g_r.$$

For very high temperatures $\rho\ (\nu_{rs}) \to \infty$, so that the expression in the bracket must $\to 0$ as $t \to \infty$,

$$\therefore\ a_{sr}g_r = a_{rs}g_s,$$

$$\therefore\ \rho\ (\nu_{rs}) = \frac{b_{rs}}{a_{sr}} \cdot \frac{1}{e^{(E_r - E_s)/Rt} - 1} = \frac{b_{rs}}{a_{sr}} \frac{1}{e^{h\nu_{rs}/Rt} - 1},$$

using

$$E_r - E_s = h\nu_{rs}.$$

Thus the density of energy of frequency ν is

$$\rho\ (\nu) = \frac{C}{e^{h\nu/Rt} - 1}.$$

[1] 'The Dynamical Theory of Gases,' by J. H. JEANS, chap. v.

Since $|\rho(\nu)\,d\nu| = |E_\lambda d\lambda|$ (reverting to the notation of § 2),

$$\rho(\nu)\,.\,\frac{c}{\lambda^2} = E_\lambda, \text{ since } \nu = \frac{c}{\lambda},$$

$$\therefore\ E_\lambda = \frac{A}{\lambda^2\,(e^{h\nu/Rt} - 1)},$$

where A is independent of t.

Einstein determined A by assuming that as $t \rightarrow \infty$, the formula reduces to the classical formula

$$E_\lambda = 8\pi Rt/\lambda^4 \text{ (§ 9)},$$

$$\therefore\ \frac{A}{\lambda^2}\frac{R}{h\nu} = 8\pi R/\lambda^4,$$

or

$$A = \frac{8\pi h\nu}{\lambda^2},$$

$$\therefore\ E_\lambda = \frac{8\pi h\nu}{\lambda^4\,(e^{h\nu/Rt} - 1)},$$

which is Planck's formula.

Eddington[1] has recently deduced Planck's formula by an extension of Einstein's method but without the use of Boltzmann's formula $N = ge^{-E/Rt}$; in fact he obtains the Boltzmann formula as a deduction from Einstein's theory.

[1] A. S. Eddington, Phil. Mag. 50, p. 803, 1925.

CHAPTER IV

THE ADIABATIC PRINCIPLE OF EHRENFEST; THE STATIONARY STATES OF A PERIODIC SYSTEM

27. *The stationary states of a periodic orbit.* The Nicholson-Bohr condition for the stationary states of the hydrogen atom was that the angular momentum should be $\frac{nh}{2\pi}$, or $mav = \frac{nh}{2\pi}$, in the notation used.

This is $$(2\pi a)\, mv = nh,$$

or $\int mv\,ds = nh$, integrated round the orbit (a circle),

or $\int_0^\sigma mv^2 dt = nh$, where σ is the periodic time $\left(= \frac{1}{\omega}\right)$.

If T is the kinetic energy, this is $2\int_0^\sigma T\,dt = nh$.

If $\bar{T}$ is the mean kinetic energy in a period,

$$\bar{T} = \frac{\int_0^\sigma T\,dt}{\int_0^\sigma dt} = \frac{1}{\sigma}\int_0^\sigma T\,dt.$$

Therefore the stationary states are given by

$$2\sigma\bar{T} = nh \text{ or } \frac{2\bar{T}}{\omega} = nh,$$

where ω is the frequency of revolution.

It is significant that the quantity which is equated to nh in the quantum condition, namely $2\bar{T}/\omega$, is already known as an 'adiabatic invariant' for any system with a single period, as was proved by Boltzmann[1] in 1876 when basing the second

[1] L. Boltzmann, 'Vorlesungen über Mechanik,' 2, § 48.

law of thermodynamics upon statistical mechanics. If the motion of a dynamical system is varied infinitely slowly by gradually altering the external field of force or slowly modifying the structure of the system, a quantity which remains constant during the change is an 'adiabatic invariant.'

Assuming for the moment, with Ehrenfest, that an adiabatic invariant is a proper magnitude to equate to nh, then for *any* periodic system the quantum condition is $2\bar{T}/\omega = nh$.

Let the system have coordinates $q_1 \dots q_n$. The corresponding momenta are $p_1 \dots p_n$, where $p_1 = \dfrac{\partial T}{\partial \dot{q}_1}$, etc., T being the kinetic energy.

Therefore $2T$ being homogeneous of degree 2 in the $\dot{q}$'s is equal to $\Sigma \dfrac{\partial T}{\partial \dot{q}} \dot{q} = \Sigma p\dot{q}$,

$$\therefore \quad 2\bar{T} = \frac{\int_0^\sigma 2T\,dt}{\sigma} = \frac{\Sigma \int_0^\sigma p\dot{q}\,dt}{\sigma} = \omega\Sigma \int_0^\sigma p\dot{q}\,dt,$$

$$\therefore \quad \frac{2\bar{T}}{\omega} = \int_0^\sigma \Sigma p\dot{q}\,dt.$$

Thus the quantum condition for a periodic orbit in general would on this assumption be

$$\int_0^\sigma \Sigma p\dot{q}\,dt = nh,$$

or

$$\int \Sigma p\,dq = nh,$$

where the integral is taken over a period of the motion.

28. *The adiabatic principle of Ehrenfest*. The stationary states in the simple system considered up to now have been determined by equating a certain mechanical entity, the angular momentum, to an integral multiple of Planck's constant, h. The assumption has been justified by its explanation with great accuracy of the hydrogen spectrum. The question now arises, what mechanical entity is to be equated to nh

for more complex systems than that of the hydrogen atom; what are the *generalised* quantum conditions to be? It is to Ehrenfest[1] that we owe the guiding principle. He arrived at that, by not limiting himself to the consideration of an isolated atomic system, but by considering the effect of external forces on the system. If those forces are sudden and violent, it is known that the 'entity' to be equated to nh changes abruptly owing to a quantum leap from n_1 to n_2, so that if these forces are slowly and gradually applied it is presumed that a quantum change will not occur and the 'entity' will not change, because it can only change by a leap.

Ehrenfest supposes that in the latter case the laws of ordinary mechanics can be applied, so that the 'entity' which does not change under the influence of the slowly changing external forces must be an 'adiabatic invariant' of the classical theory. This is the 'adiabatic principle' of Ehrenfest, and it requires that only adiabatic invariants are to be equated to nh in order to determine the stationary states.

29. *Illustration from the simple pendulum.* At the Solvay Congress[2] at Brussels in 1911, Lorentz recalled a conversation which he had had with Einstein in which the question of the change of the motion of a simple pendulum, the thread of which was held between the fingers and shortened by drawing it up between them, was discussed. Einstein replied that if the length of the pendulum is changed infinitely slowly, the ratio of the kinetic energy of the pendulum to its frequency would remain constant. This illustrates the principle stated above for periodic systems in general, and for the pendulum can be worked out as follows.

If l is the length of the thread and θ its inclination to the vertical, the frequency ω is given by

$$\frac{1}{\omega} = 2\pi\sqrt{\frac{l}{g}} \text{ or } 4\pi^2\omega^2 l = g \quad \ldots\ldots\ldots\ldots(1),$$

[1] P. Ehrenfest, Kersl. Akad. Amsterdam, **16**, p. 591, 1914; Ann. d. Phys. **51**, p. 327, 1916; Phil. Mag. **33**, p. 500, 1917.

[2] La Théorie du Rayonnement et les Quanta,' a report of the 1911 Solvay Congress, Gauthier-Villars, p. 450, 1912.

the tension R of the thread is given by $ml\dot{\theta}^2 = R - mg\cos\theta$, and the angle $\theta = A\cos(2\pi\omega t)$.

If the length of the thread is altered from l to $l + dl$, the work done by the outside agency is $-Rdl$, and if the change is so slow that the time taken is a large number of periods, the work done is $-\overline{R}dl$, where the bar denotes the mean value taken for a large number of periods.

The work done is thus

$$-dl\,[m\overline{l\dot{\theta}^2} + mg\,\overline{\cos\theta}].$$

Now $\overline{l\dot{\theta}^2} = \overline{lA^2 4\pi^2\omega^2\sin^2 2\pi\omega t} = lA^2\,2\pi^2\omega^2 = \frac{1}{2}gA^2$, using (1), and

$$\overline{\cos\theta} = 1 - \frac{\overline{\theta^2}}{2} = 1 - \tfrac{1}{2}.\overline{A^2\cos^2(2\pi\omega t)} = 1 - \frac{A^2}{4},$$

therefore the work done is

$$-mg\,dl\left[\frac{A^2}{2} + \left(1 - \frac{A^2}{4}\right)\right] = -mgdl\left(1 + \frac{A^2}{4}\right) \quad\text{...(2).}$$

If T, V are the kinetic and potential energies of the pendulum,

$$T = \tfrac{1}{2}ml^2\dot{\theta}^2 \text{ and } V = C - mgl\cos\theta,$$

$$\therefore\ \overline{T} = \tfrac{1}{2}ml^2A^2 4\pi^2\overline{\omega^2\sin^2(2\pi\omega t)} = \tfrac{1}{4}ml^2A^2 4\pi^2\omega^2 = \frac{mglA^2}{4},$$

using (1), and

$$\overline{V} = C - mgl\,\{1 - \tfrac{1}{2}A^2\overline{\cos^2(2\pi\omega t)}\} = C - mgl + \frac{mglA^2}{4}.$$

Since the work done is the increase of $\overline{T} + \overline{V}$, we have

$$-mgdl\left(1 + \frac{A^2}{4}\right) = d\left[C - mgl + mgl\frac{A^2}{2}\right],$$

$$\therefore\ -A^2dl = 2d\,(lA^2) = 2dl.A^2 + 4lAdA,$$

$$\therefore\ 3Adl + 4ldA = 0,$$

whence $l^3A^4 = \text{constant} = k$, suppose ....(3).

$$\text{Hence}\quad \frac{\overline{T}}{\omega} = \frac{mglA^2}{4}.2\pi\sqrt{\frac{l}{g}}$$

$$= \frac{\pi m}{2}\sqrt{gl^3A^4} = \frac{\pi m}{2}\sqrt{gk}, \text{ from (3),}$$

and is constant during the change.

30. *Elliptic orbits of an electron.* If the nucleus is at the focus, the force per unit mass on the electron is $\frac{Ze^2}{mr^2}$ (where Ze is the nuclear charge) and is $\frac{\mu}{r^2}$, where $\mu = Ze^2/m$.

The polar equation of the orbit is $\frac{l}{r} = 1 + \epsilon \cos \theta$, where l is the semi-latus rectum ($= b^2/a$) and ϵ is the eccentricity. If v is the velocity,

$$v^2 = \mu \left(\frac{2}{r} - \frac{1}{a}\right) \text{ and } r^2\dot{\theta} = \text{constant} = \sqrt{\mu l} \quad \ldots(1).$$

The frequency of revolution $(\omega) = 1/\text{period} = \frac{1}{2\pi}\sqrt{\frac{\mu}{a^3}}$.

All these results are well known from dynamics.

If W is the negative energy

$$-W = \tfrac{1}{2}mv^2 - \frac{Ze^2}{r},$$

$$\therefore \quad W = \frac{Ze^2}{r} - \tfrac{1}{2}mv^2 = \frac{Ze^2}{r} - \frac{m}{2}\mu\left(\frac{2}{r} - \frac{1}{a}\right)$$

$$= \frac{Ze^2}{r} - \frac{m}{2}\left(\frac{Ze^2}{m}\right)\left(\frac{2}{r} - \frac{1}{a}\right),$$

or

$$W = \frac{Ze^2}{2a}.$$

The quantum condition is $\frac{2\bar{T}}{\omega} = nh$ or

$$2\int_0^\sigma T\,dt = nh, \text{ where } \sigma = \text{the period} = \frac{1}{\omega}.$$

$$\therefore \quad \int_0^\sigma mv^2\,dt = nh,$$

$$\int_0^\sigma m\mu\left(\frac{2}{r} - \frac{1}{a}\right)dt = nh,$$

$$Ze^2.\int_0^\sigma \left(\frac{2}{r} - \frac{1}{a}\right)dt = nh,$$

$$\therefore \quad Ze^2\int_0^{2\pi}\left(\frac{2}{r} - \frac{1}{a}\right)\frac{d\theta}{\dot{\theta}} = nh,$$

$$Ze^2\int_0^{2\pi}\left(\frac{2}{r}-\frac{1}{a}\right)\frac{r^2d\theta}{\sqrt{\mu l}}=nh, \text{ using (1)},$$

$$\frac{Ze^2}{\sqrt{\mu l}}\left[\int_0^{2\pi}2r\,d\theta-\frac{1}{a}\int_0^{2\pi}r^2\,d\theta\right]=nh.$$

Now

$$\int_0^{2\pi}r\,d\theta=l\int_0^{2\pi}\frac{d\theta}{1+\epsilon\cos\theta}=\frac{2\pi l}{\sqrt{1-\epsilon^2}}=2\pi\frac{b^2}{a}\Big/\frac{b}{a}=2\pi b,$$

and $\int_0^{2\pi}r^2d\theta =$ twice the area of the ellipse $=2\pi ab$,

$$\therefore\ \frac{Ze^2}{\sqrt{\mu l}}\left[4\pi b-\frac{1}{a}(2\pi ab)\right]=nh,$$

$$\therefore\ 2\pi be^2Z\Big/\sqrt{\frac{\mu b^2}{a}}=nh,$$

$$\therefore\ \frac{2\pi Ze^2\sqrt{a}}{\sqrt{\frac{Ze^2}{m}}}=nh,$$

or

$$4\pi^2Ze^2am=n^2h^2,$$

$$\therefore\ a=\frac{n^2h^2}{4\pi^2Ze^2m},$$

and

$$W=\frac{Ze^2}{2a}=\frac{2\pi^2me^4Z^2}{n^2h^2}.$$

The frequency in the orbit

$$\omega=\frac{1}{2\pi}\sqrt{\frac{\mu}{a^3}}=\frac{1}{2\pi}\frac{e\sqrt{Z}}{\sqrt{m}}\cdot\left(\frac{8\pi^3m^{\frac{3}{2}}e^3Z^{\frac{3}{2}}}{n^3h^3}\right)=\frac{4\pi^2me^4Z^2}{n^3h^3}.$$

These results are identical with those for a circular orbit of radius a. Thus the energy of the orbit is determined by the length of its major axis and is independent of the eccentricity. The orbits of quantum number n are infinite in number; all have the same major axis, but the minor axis may have any value.

31. *The equation $\delta E = \omega \delta I$ for any simply periodic system*[1].

Let $I = \int_0^\sigma \Sigma p\dot{q}\,dt$ and consider a small variation of the motion which leaves the system still simply periodic.

For the variation of I we have

$$\delta I = \int_0^\sigma \Sigma\,(\dot{q}\,\delta p + p\,\delta\dot{q})\,dt + \Big[\Sigma p\dot{q}\,\delta t\Big]_0^\sigma,$$

where the last term is due to the variation in the period σ.

Since $\int_0^\sigma p\,\delta\dot{q}\,dt = \int_0^\sigma p\,d\,(\delta q) = \Big[p\,\delta q\Big]_0^\sigma - \int_0^\sigma \delta q\,\dot{p}\,dt,$

we have

$$\delta I = \int_0^\sigma \Sigma\,(\dot{q}\,\delta p - \dot{p}\,\delta q)\,dt + \Big[\Sigma p\,(\dot{q}\,\delta t + \delta q)\Big]_0^\sigma.$$

The last term vanishes owing to the periodicity, so that

$$\delta I = \int_0^\sigma \Sigma\,(\dot{q}\,\delta p - \dot{p}\,\delta q)\,dt.$$

Using Hamilton's equations (§ 38),

$$\frac{\partial H}{\partial p} = \dot{q}, \quad \frac{\partial H}{\partial q} = -\,\dot{p},$$

it follows that $\delta I = \int_0^\sigma \Sigma\left(\frac{\partial H}{\partial p}\,\delta p + \frac{\partial H}{\partial q}\,\delta q\right) dt$

$$= \int_0^\sigma \delta H\,.\,dt.$$

Also H is the total energy, $= E$ (§ 39).

$$\therefore\ \delta I = \int_0^\sigma \delta E\,.\,dt.$$

If the varied motion is one corresponding to slightly different initial conditions, then δE is constant, so that

$$\delta I = (\delta E)\,\sigma, \text{ since } \sigma = \frac{1}{\omega},$$

it follows that $\qquad \delta E = \omega\,\delta I.$

[1] N. Bohr, 'On the Quantum Theory of Line Spectra,' D. Kgl. Danske Vidensk. Skrifter, 8. iv. 1, p. 10, 1918. This will be referred to afterwards as 'Q.L.S.'

32. *Bohr's theory for a simply periodic system.* Let

$$I = \int_0^\sigma \Sigma p\dot{q}\,dt \text{ and } A = \Sigma p\dot{q}.$$

If $\bar{A}$ is the mean value of A over a period

$$\bar{A} = \frac{1}{\sigma}\int_0^\sigma \Sigma p\dot{q}\,dt = \frac{I}{\sigma} = I\omega.$$

Bohr's theory is based upon the equations $\bar{A} = I\omega$ and $\delta E = \omega\delta I$.

Let the frequency ω of the system be found in terms of E. The second equation on integration determines E in terms of I, except for an arbitrary constant, which the first equation determines.

The quantum condition $I = nh$ then determines E in terms of n.

For example, in the Keplerian orbit, with nuclear charge e,

$$W = \frac{e^2}{2a} \text{ and } \omega = \frac{1}{2\pi}\sqrt{\frac{e^2}{ma^3}} \quad (\S\ 30).$$

From these by elimination of a it follows that

$$\omega^2 = \frac{2W^3}{\pi^2 me^4}.$$

But $E = -W$, so that

$$\delta W = -\delta E = -\omega\delta I = -\delta I\sqrt{\frac{2W^3}{\pi^2 me^4}}.$$

Integrating $$\frac{2}{\sqrt{W}} = I\sqrt{\frac{2}{\pi^2 me^4}} + C,$$

where C is a constant.

From the equation $\bar{A} = I\omega$, $I \to 0$ as $\omega \to \infty$, or as $W \to \infty$,

$$\therefore\ C = 0,$$

$$\therefore\ \frac{2\pi^2 me^4}{I^2} = W.$$

The condition $I = nh$ then determines the negative energy W in the stationary states as

$$W = \frac{2\pi^2 me^4}{h^2 n^2}.$$

Also it follows that $\omega = \frac{4\pi^2 e^4 m}{I^3}$.

The correspondence principle is then deduced by considering a transition from n_1 to n_2.

For $\nu = \frac{1}{h}[E_1 - E_2] = \frac{1}{h}\Sigma_2^1 \delta E = \frac{1}{h}\Sigma_2^1 \omega\delta I.$

If n_1 and n_2 are large compared with their difference, the motions in the two states differ little from one another, so that ω may be taken approximately constant in the summation, and

$$\nu = \frac{1}{h}\omega(I_1 - I_2)$$
$$= \omega(n_1 - n_2), \text{ since } I = nh.$$

Thus in the limit, for large quantum numbers, the frequency of the radiation emitted will be the same as it would on the classical theory (§ 23) and the conclusions of that paragraph follow.

CHAPTER V

GENERAL DYNAMICAL THEORY

33. *Newtonian theory.* The kinetic energy T of a particle is $\frac{1}{2}m(\dot{x}^2 + \dot{y}^2 + \dot{z}^2)$, where x, y, z are the rectangular coordinates of the particle and m its mass. The momentum p_x parallel to Ox is $m\dot{x}$ and is therefore $\frac{\partial T}{\partial \dot{x}}$, so that

$$p_x = \frac{\partial T}{\partial \dot{x}}, \text{ etc.} \quad \text{.....................(1).}$$

The equations of motion are $\dot{p}_x = X$, etc., where X, Y, Z are the external forces on the particle; if these forces have a potential energy V,

$$X = -\frac{\partial V}{\partial x}, \text{ etc.,}$$

so that

$$\dot{p}_x = -\frac{\partial V}{\partial x}, \text{ etc.} \quad \text{.....................(2).}$$

Also multiplying (2) by $\dot{x}$, etc., and adding

$$(\dot{p}_x\dot{x} + +) = -\frac{\partial V}{\partial x}\dot{x} - - - = -\frac{dV}{dt},$$

$$\therefore \ (m\ddot{x}\,.\,\dot{x} + +) = -\frac{dV}{dt}.$$

Therefore integrating,

$$\tfrac{1}{2}m(\dot{x}^2 + +) + V = \text{constant}$$

or $\qquad T + V = \text{constant}$, (the equation of energy).

34. *The variation principle.* Let $L = T - V$ and consider $\int_{t_1}^{t_2} L\,dt$ for the actual path of the particle between times t_1 and t_2. Let δ denote an arbitrary small variation of the coordinates and velocities by means of which the particle can pass from its position at time t_1 to its position at time t_2 but by a path slightly different from its actual path.

Then $\delta\int_{t_1}^{t_2} L\,dt = \delta\int_{t_1}^{t_2}(T - V)\,dt = \int_{t_1}^{t_2}(\delta T - \delta V)\,dt$

$$= \int_{t_1}^{t_2}\left\{\left(\frac{\partial T}{\partial \dot{x}}\,\delta\dot{x} + +\right) - \delta V\right\} dt$$

$$= \int_{t_1}^{t_2}\{(p_x\delta\dot{x} + +) - \delta V\}\,dt.$$

But $\int_{t_1}^{t_2} p_x\delta\dot{x}\,dt = \int_{t_1}^{t_2} p_x d\,(\delta x) = (p_x\delta x)_{t_1}^{t_2} - \int_{t_1}^{t_2}\delta x\,\dot{p}_x dt$

and δx vanishes at times t_1, t_2, since the varied state is also the actual one at those times, so that

$$\int_{t_1}^{t_2} p_x\delta\dot{x}\,dt = -\int_{t_1}^{t_2}\delta x\,\dot{p}_x dt,$$

$$\therefore\ \delta\int_{t_1}^{t_2} L\,dt = \int_{t_1}^{t_2} - (\delta x\,\dot{p}_x + +)\,dt - \int_{t_1}^{t_2}\left(\frac{\partial V}{\partial x}\,\delta x + +\right) dt$$

$$= -\int_{t_1}^{t_2}\left[\left(\dot{p}_x + \frac{\partial V}{\partial x}\right)\delta x + +\right] dt = 0, \text{ by equations (2).}$$

Thus $\delta\int_{t_1}^{t_2} L\,dt = 0$, where $L = T - V$.

This is the 'variation principle,' and its essential advantage is its independence of any coordinate system.

35. *Lagrange's equations.* If the position of the particle can be determined by s independent coordinates $q_1 \ldots q_s$ so that T is a function of q, $\dot{q}$, where q means any of the q's, and V is a function of q, then L is a function of q, $\dot{q}$.

The variation principle $\delta\int L\,dt = 0$, where the limits of the integral are understood, leads to

$$0 = \int \delta L\,dt = \int \Sigma\left(\frac{\partial L}{\partial q}\,\delta q + \frac{\partial L}{\partial \dot{q}}\,\delta\dot{q}\right) dt.$$

Writing $$p = \frac{\partial T}{\partial \dot{q}} = \frac{\partial L}{\partial \dot{q}},$$

this becomes $$0 = \int \Sigma\left(\frac{\partial L}{\partial q}\,\delta q + p\,\delta\dot{q}\right) dt.$$

But $$\int p\delta\dot{q}\,dt = \int pd\,(\delta q) = [p\delta q] - \int \delta q\,\dot{p}\,\delta t,$$

and the term in brackets vanishes as before, so that the equation becomes

$$0 = \int \Sigma\left(\frac{\partial L}{\partial q} - \dot{p}\right)\delta q\,dt.$$

The δq's are arbitrary, so that $\frac{\partial L}{\partial q} = \dot{p}$.

Hence we have finally the equations

$$\left.\begin{aligned} p &= \frac{\partial L}{\partial \dot{q}} \\ \dot{p} &= \frac{\partial L}{\partial q} \end{aligned}\right\} \qquad \text{........................(3).}$$

These are Lagrange's equations, the usual form being the result of eliminating p, which is

$$\frac{d}{dt}\left(\frac{\partial L}{\partial \dot{q}}\right) = \frac{\partial L}{\partial q}.$$

36. *Relativity theory.* The Einstein equations are $\dot{p}_x = X$, etc., where $p_x = m\dot{x}$, but now m is no longer constant, but is given by

$$m = \frac{m_0}{\sqrt{1 - \frac{v^2}{c^2}}},$$

on the theory of relativity[1], where v is the velocity of m and c is the velocity of light. This formula expresses the variation of mass with velocity, and m_0 is the 'rest-mass,' i.e. the mass when $v = 0$. Write $\beta = \frac{v}{c}$.

Let $p_x = \frac{\partial T'}{\partial \dot{x}}$, where T' is a new function, to be found.

Then $$\frac{\partial T'}{\partial \dot{x}} = \frac{m_0 \dot{x}}{\sqrt{1 - \beta^2}},$$

$$dT' = \frac{\partial T'}{\partial \dot{x}}\, d(\dot{x}) + + = m_0 . \frac{\dot{x}\, d(\dot{x}) + +}{\sqrt{1 - \beta^2}},$$

[1] 'Space, Time, and Gravitation,' by A. S. Eddington, p. 145, Cambridge, 1920.

and
$$\beta^2 = \frac{v^2}{c^2} = \frac{\dot{x}^2 + +}{c^2},$$

so that
$$c^2 \beta d\beta = \dot{x} d(\dot{x}) + +.$$

$$\therefore \; dT' = \frac{m_0 c^2 \beta d\beta}{\sqrt{1 - \beta^2}},$$

$$\therefore \; T' = - m_0 c^2 \sqrt{1 - \beta^2} + A.$$

Now when $\frac{v}{c}$ is small, relativity mechanics merges into Newtonian, and there $p_x = \frac{\partial T}{\partial \dot{x}}$, where T is the Newtonian kinetic energy, $\frac{1}{2} m_0 v^2$. So that T' should $\to \frac{1}{2} m_0 v^2$ as $\beta \to 0$. This fixes the constant A; for

$$\tfrac{1}{2} m_0 v^2 = - m_0 c^2 \left(1 - \frac{\beta^2}{2}\right) + A, \text{ for small values of } \frac{v}{c},$$

$$= - m_0 c^2 \left(1 - \frac{v^2}{2c^2}\right) + A.$$

$$\therefore \; A = m_0 c^2,$$

and
$$T' = m_0 c^2 [1 - \sqrt{1 - \beta^2}].$$

Thus in relativity mechanics we have the equations

$$p_x = \frac{\partial T'}{\partial \dot{x}} \text{ and } \dot{p}_x = - \frac{\partial V}{\partial x}$$

corresponding precisely to equations (1) and (2) of § 33.

Therefore if now we write $L = T' - V$, the variation principle follows, and from it Lagrange's equations

$$p = \frac{\partial L}{\partial \dot{q}} \text{ and } \dot{p} = \frac{\partial L}{\partial q},$$

of the same form as before.

The new L is no longer the difference of the kinetic and potential energies but is

$$m_0 c^2 [1 - \sqrt{1 - \beta^2}] - V,$$

and the Lagrangian equations have the same form as in Newtonian mechanics.

The equation of energy is deduced as in the Newtonian

case by multiplying the equations $\dot{p}_x = -\frac{\partial V}{\partial x}$, etc. by $\dot{x}$, etc. and adding.

It follows that $$(\dot{p}_x \dot{x} + +) = -\frac{dV}{dt}$$
as before, or
$$m_0 \left[\frac{d}{dt}\left(\frac{\dot{x}}{\sqrt{1-\beta^2}}\right).\dot{x} + +\right] = -\frac{dV}{dt},$$
$$m_0 \left[\frac{\dot{x}\ddot{x} + +}{\sqrt{1-\beta^2}} + \frac{(\dot{x}^2 + +)\,\beta\dot{\beta}}{(1-\beta^2)^{\frac{3}{2}}}\right] = -\frac{dV}{dt}.$$
Since $$\beta^2 c^2 = v^2 = \dot{x}^2 + \dot{y}^2 + \dot{z}^2,$$
then $$\beta\dot{\beta} c^2 = \dot{x}\ddot{x} + +,$$
and the equation becomes
$$\frac{m_0 \beta\dot{\beta}}{(1-\beta^2)^{\frac{3}{2}}}\{c^2(1-\beta^2) + v^2\} + \frac{dV}{dt} = 0,$$
or $$\frac{m_0 c^2 \beta\dot{\beta}}{(1-\beta^2)^{\frac{3}{2}}} + \frac{dV}{dt} = 0,$$
whence by integration,
$$\frac{m_0 c^2}{\sqrt{1-\beta^2}} + V = \text{constant}.$$
Thus the kinetic energy must be
$$\frac{m_0 c^2}{\sqrt{1-\beta^2}} + C,$$
and when $v = 0$ it should also be zero, so that
$$m_0 c^2 + C = 0.$$
Therefore the kinetic energy
$$T = m_0 c^2 \left[\frac{1}{\sqrt{1-\beta^2}} - 1\right].$$
When β is very small this reduces to the usual $\frac{1}{2} m_0 v^2$.

37. *The dynamics of an electron in a magnetic field.* Let the forces on the electron due to the magnetic field be X, Y, Z; they are *not derivable from a potential*. If V is the potential of the other forces acting on the electron, the equations $\dot{p}_x = -\frac{\partial V}{\partial x}$, etc. become $\dot{p}_x = -\frac{\partial V}{\partial x} + X$, etc.

If relativity is neglected,

$$p_x = \frac{\partial T}{\partial \dot{x}} = \frac{\partial}{\partial \dot{x}}(T - V), \text{ since } V \text{ does not contain } \dot{x},$$

and $$\dot{p}_x = -\frac{\partial V}{\partial x} + X = -\frac{\partial}{\partial x}(T - V) + X,$$

so that $$\frac{d}{dt}\left\{\frac{\partial}{\partial \dot{x}}(T - V)\right\} - \frac{\partial}{\partial x}(T - V) = X.$$

Now let M be a function such that

$$\frac{d}{dt}\left(\frac{\partial M}{\partial \dot{x}}\right) - \frac{\partial M}{\partial x} = X;$$

it follows that

$$\frac{d}{dt}\left\{\frac{\partial}{\partial \dot{x}}(T - V - M)\right\} - \frac{\partial}{\partial x}(T - V - M) = 0.$$

Thus the Lagrangian equations have their usual form with the Lagrangian function

$$L = T - V - M.$$

So, if relativity is taken into account,

$$L = T' - V - M.$$

If the magnetic force is α, β, γ, it is known from electromagnetic theory that the mechanical force X, Y, Z on a moving electron is

$$X = \frac{(-e)}{c}(\dot{y}\gamma - \dot{z}\beta), \text{ etc.}$$

Writing $\alpha = \frac{\partial H}{\partial y} - \frac{\partial G}{\partial z}$, etc., where F, G, H is the vector potential, it follows that

$$X = -\frac{e}{c}\left\{\dot{y}\left(\frac{\partial G}{\partial x} - \frac{\partial F}{\partial y}\right) - \dot{z}\left(\frac{\partial F}{\partial z} - \frac{\partial H}{\partial x}\right)\right\}$$

$$= -\frac{e}{c}\left\{\left(\dot{x}\frac{\partial F}{\partial x} + \dot{y}\frac{\partial G}{\partial x} + \dot{z}\frac{\partial H}{\partial x}\right) - \left(\dot{x}\frac{\partial F}{\partial x} + \dot{y}\frac{\partial F}{\partial y} + \dot{z}\frac{\partial F}{\partial z}\right)\right\},$$

or $$X = -\frac{e}{c}\left\{\frac{\partial}{\partial x}(\dot{x}F + \dot{y}G + \dot{z}H) - \frac{dF}{dt}\right\},$$

$$\therefore \quad \frac{d}{dt}\left(\frac{\partial M}{\partial \dot{x}}\right) - \frac{\partial M}{\partial x} = X = -\frac{e}{c}\left\{\frac{\partial}{\partial x}(\dot{x}F + +) - \frac{dF}{dt}\right\}.$$

Thus $$M = \frac{e}{c}(\dot{x}F + \dot{y}G + \dot{z}H)$$

satisfies this equation, and

$$L = T - V - M = T - V - \frac{e}{c}(\dot{x}F + \dot{y}G + \dot{z}H).$$

38. *Hamilton's equations.* Let $H = \Sigma p\dot{q} - L$. The $\dot{q}$'s can be expressed in terms of p, q by solving the equations $p = \frac{\partial L}{\partial \dot{q}}$; substituting for them in H, H can be expressed as a function of p, q.

Then if δ is an arbitrary small variation of p, q, the resulting change in H is given by

$$\begin{aligned} \delta H &= \Sigma(\dot{q}\delta p + p\delta\dot{q}) - \delta L \\ &= \Sigma[\dot{q}\delta p + p\delta\dot{q}] - \Sigma\left(\frac{\partial L}{\partial q}\delta q + \frac{\partial L}{\partial \dot{q}}\delta\dot{q}\right) \\ &= \Sigma(\dot{q}\delta p + p\delta\dot{q} - \dot{p}\delta q - p\delta\dot{q}), \end{aligned}$$

using Lagrange's equations $p = \frac{\partial L}{\partial \dot{q}}$, $\dot{p} = \frac{\partial L}{\partial q}$.

$$\therefore \quad \delta H = \Sigma(\dot{q}\delta p - \dot{p}\delta q).$$

Hence $$\frac{\partial H}{\partial p} = \dot{q}, \quad \frac{\partial H}{\partial q} = -\dot{p}.$$

These are the 'canonical' equations of Hamilton.

39. On the *Newtonian* theory, T, in the applications to be made later, will be quadratic in the $\dot{q}$'s, as the time will not occur explicitly in the geometry of the system, so that

$$\Sigma\dot{q}\frac{\partial T}{\partial \dot{q}} = 2T.$$

But

$$H = \Sigma p\dot{q} - L = \Sigma \dot{q}\frac{\partial T}{\partial \dot{q}} - L = 2T - L = 2T - (T - V) = T + V,$$

so that H is the total energy of the motion.

On the *relativistic* theory, $L = T' - V$, where

$$T' = m_0 c^2 (1 - \sqrt{1 - \beta^2}),$$

and the kinetic energy

$$T = m_0 c^2 \left(\frac{1}{\sqrt{1 - \beta^2}} - 1\right).$$

Hence $H = \Sigma p\dot{q} - L = \Sigma \frac{\partial L}{\partial \dot{q}}\dot{q} - L = \Sigma \frac{\partial T'}{\partial \dot{q}}\dot{q} - L$

$$= \Sigma \frac{m_0 c^2}{\sqrt{1 - \beta^2}} \beta \frac{\partial \beta}{\partial \dot{q}} \dot{q} - (T' - V).$$

But $$\beta^2 = \frac{v^2}{c^2} = \frac{\dot{x}^2 + \dot{y}^2 + \dot{z}^2}{c^2},$$

and is therefore homogeneous of degree 2 in the $\dot{q}$'s, so that

$$\Sigma \dot{q} \frac{\partial}{\partial \dot{q}} (\beta^2) = 2\beta^2,$$

or $$\Sigma \beta \frac{\partial \beta}{\partial \dot{q}} \dot{q} = \beta^2,$$

$$\therefore\ H = \frac{m_0 c^2 \beta^2}{\sqrt{1 - \beta^2}} - T' + V$$

$$= \frac{m_0 v^2}{\sqrt{1 - \beta^2}} - \Sigma m_0 c^2 (1 - \sqrt{1 - \beta^2}) + V$$

$$= \frac{m_0}{\sqrt{1 - \beta^2}} [v^2 - c^2\sqrt{1 - \beta^2} + c^2 - v^2] + V$$

$$= m_0 c^2 \left[\frac{1}{\sqrt{1 - \beta^2}} - 1\right] + V$$

$$= T + V.$$

Thus H is still the total energy of the system.

CHAPTER VI

CONTACT TRANSFORMATIONS; THE HAMILTON-JACOBI DIFFERENTIAL EQUATION

40. *Contact transformations.* The variation principle which leads to Hamilton's canonical form is $\delta \int L dt = 0$. H is a function of p, q and will now be used as a functional symbol indicating this fact and written $H(p, q)$.

Since $H(p, q) = \Sigma p\dot{q} - L$, the variation principle is

$$\delta \int [\Sigma p\dot{q} - H(p, q)]\, dt = 0,$$

and leads to the equations

$$\frac{\partial H}{\partial p} = \dot{q}, \quad \frac{\partial H}{\partial q} = -\dot{p}.$$

The object is now to find new variables P, Q for which the equations of motion keep their canonical form. This will be the case if the equation

$$\delta \int [\Sigma p\dot{q} - H(p, q)]\, dt = 0$$

becomes

$$\delta \int [\Sigma P\dot{Q} - K(P, Q)]\, dt = 0.$$

This is only so if the difference of the integrands is a perfect differential $\frac{dV}{dt}$ of a function V of $2s$ of the new and old variables (s being the number of each of the variables p, q, P, Q).

If we choose V to be a function of $q, Q, t \equiv F(q, Q, t)$, then

$$\begin{aligned}\Sigma p\dot{q} - H(p, q) &= \Sigma P\dot{Q} - K(P, Q) + \frac{d}{dt} F(q, Q, t)\\ &= \Sigma P\dot{Q} - K(P, Q) + \Sigma \frac{\partial F}{\partial q}\dot{q} + \Sigma \frac{\partial F}{\partial Q}\dot{Q} + \frac{\partial F}{\partial t},\end{aligned}$$

which is satisfied by

$$\left.\begin{aligned} p=\frac{\partial F}{\partial q},\quad P=-\frac{\partial F}{\partial Q},\quad K=H+\frac{\partial F}{\partial t}\ \ldots(1)\\ \text{or by}\qquad \Sigma\,(p\delta q-P\delta Q)=\delta F,\quad K=H+\frac{\partial F}{\partial t}\ \ldots(2)\end{aligned}\right\}(\alpha).$$

A transformation expressed by the first pair of equations of (1) or the first equation of (2) is a '*contact transformation.*'

The canonical equations

$$\frac{\partial H}{\partial p}=\dot q,\quad \frac{\partial H}{\partial q}=-\dot p$$

become

$$\frac{\partial K}{\partial P}=\dot Q,\quad \frac{\partial K}{\partial Q}=-\dot P,$$

where

$$K=H+\frac{\partial F}{\partial t}.$$

41. *Other forms of contact transformation.* If V is chosen of the form

$$F\,(q, P, t)-\Sigma PQ,$$

then

$$\Sigma p\dot q-H\,(p,q)=\Sigma P\dot Q-K\,(P,Q)+\frac{d}{dt}[F\,(q,P,t)-\Sigma PQ]$$

$$=-\,\Sigma Q\dot P-K\,(P,Q)+\Sigma\frac{\partial F}{\partial q}\dot q+\Sigma\frac{\partial F}{\partial P}\dot P+\frac{\partial F}{\partial t},$$

which is satisfied if

$$\left.\begin{aligned} p=\frac{\partial F}{\partial q},\quad Q=\frac{\partial F}{\partial P}\ \text{and}\ K=H+\frac{\partial F}{\partial t}\\ \text{or if}\qquad \Sigma\,(p\delta q+Q\delta P)=\delta F\ \text{and}\ K=H+\frac{\partial F}{\partial t}\end{aligned}\right\}\ldots\ldots(\beta).$$

The new equations are still

$$\frac{\partial K}{\partial P}=\dot Q,\quad \frac{\partial K}{\partial Q}=-\dot P.$$

So by using $\quad V=F\,(p,Q,t)-\Sigma PQ$

we find the transformation

$$\Sigma\,(p\delta q+P\delta Q)=-\,\delta F,\quad K=H+\frac{\partial F}{\partial t}\quad\ldots\ldots(\gamma),$$

and by using $V = F(p, P, t) - \Sigma PQ - \Sigma pq$
we obtain the transformation

$$\Sigma (q\delta p - Q\delta P) = -\delta F, \quad K = H + \frac{\partial F}{\partial t} \quad \text{......}(\delta),$$

both of which lead to

$$\frac{\partial K}{\partial P} = \dot{Q}, \quad \frac{\partial K}{\partial Q} = -\dot{P}$$

as before.

In the applications to be made later, F usually does not contain t explicitly, so that $K = H$; or the new Hamiltonian function is equal to the old one.

42. *The Hamilton-Jacobi differential equation.* Let the variables p, q be transformed to ϕ, α by the transformation $\Sigma (p\delta q + \phi\delta\alpha) = \delta S$, where S is a function of q, α and does not contain the time explicitly, so that the transformation is of type (β) of § 41, and H is the Hamiltonian function for both sets of variables.

Suppose the transformation is such that H is a function of the new variables α only; then since

$$\frac{\partial H}{\partial \alpha} = \dot{\phi}, \quad \frac{\partial H}{\partial \phi} = -\dot{\alpha},$$

it follows, since H does not contain ϕ, that $\dot{\alpha} = 0$ and the α's are constant.

Therefore $\dfrac{\partial H}{\partial \alpha}$ is constant, since H contains only α's, and writing $\dfrac{\partial H}{\partial \alpha} = \omega$, we have $\dot{\phi} = \omega$, or $\phi = \omega t + \beta$, where β is a constant and t is the time.

If, as a special case, H reduces to α_1 and is independent of $\alpha_2, \alpha_3 \ldots$, then $H(p, q) = \alpha_1$, and since $p = \dfrac{\partial S}{\partial q}$ {from the equation $\Sigma (p\delta q + \phi\delta\alpha) = \delta S$}, we obtain

$$H\left(\frac{\partial S}{\partial q}, q\right) = \alpha_1.$$

This equation to determine S is the *Hamilton-Jacobi differential equation*. Let the complete integral of this be found and let it be $S = f(\alpha_1, \alpha_2, \dots \alpha_s) + C$, where $\alpha_2 \dots \alpha_s$, C are the s arbitrary constants of the integral, s being the number of coordinates q. C may be dropped, as in the applications, only differential coefficients of S will be required.

Then this function S used in $\Sigma(p\delta q + \phi\delta\alpha) = \delta S$ will transform the Hamiltonian function into α_1.

But since $\omega = \frac{\partial H}{\partial \alpha}$, we have in this case $\omega_1 = \frac{\partial H}{\partial \alpha_1} = 1$, $\omega_2 = \frac{\partial H}{\partial \alpha_2} = 0$, etc., or $\omega_1 = 1$, $\omega_2 = 0$, $\omega_3 = 0$, etc.

Therefore since $\phi = \omega t + \beta$, we obtain $\phi_1 = t + \beta_1$, $\phi_2 = \beta_2$, $\phi_3 = \beta_3, \dots$, or since from $\Sigma(p\delta q + \phi\delta\alpha) = \delta S$, it follows that $\phi = \frac{\partial S}{\partial \alpha}$, we have

$$\frac{\partial S}{\partial \alpha_1} = t + \beta_1, \quad \frac{\partial S}{\partial \alpha_2} = \beta_2, \quad \frac{\partial S}{\partial \alpha_3} = \beta_3, \text{ etc. } \quad \dots\dots(1).$$

These satisfy the transformed Hamiltonian equations and therefore the original ones, and are the solution of the dynamical problem.

The equations $\frac{\partial S}{\partial \alpha_2} = \beta_2$, $\frac{\partial S}{\partial \alpha_3} = \beta_3, \dots$ are $(s-1)$ equations connecting $q_1, q_2 \dots q_s$ and the constants $\alpha_1 \dots \alpha_s$, $\beta_2 \dots \beta_s$, which give the form of the orbit; the first equation $\frac{\partial S}{\partial \alpha_1} = t + \beta_1$ connects the time with the position in the orbit.

43. *Solution of the Hamilton-Jacobi equation by separation of variables.* If orthogonal coordinates are used, the Hamiltonian function $H(p, q)$ contains the p's as squares only, both in Newtonian and relativity mechanics. Thus the Hamilton-Jacobi equation $H\left(\frac{\partial S}{\partial q}, q\right) = \alpha_1$ contains the $\frac{\partial S}{\partial q}$'s only as squares.

It may happen that it is possible to dissect the equation into parts each of the form $\left(\frac{\partial S}{\partial q_k}\right)^2 = f_k(q_k, \alpha_1 \dots \alpha_s)$, such that if each part is satisfied, the Hamilton-Jacobi equation as a whole is satisfied. Any one part contains only one of the coordinates, and each part may contain any of the constants $\alpha_1 \dots \alpha_s$, where α_1 is the constant occurring in the original equation and $\alpha_2 \dots \alpha_s$ are arbitrary constants. If this process can be effected, the variables are 'separable.'

The complete integral of the Hamilton-Jacobi equation is then given by integrating

$$dS = \frac{\partial S}{\partial q_1} dq_1 + \frac{\partial S}{\partial q_2} dq_2 + \dots$$

and is $S = \int \frac{\partial S}{\partial q_1} dq_1 + \dots$

$$= \int \sqrt{f_1(q_1, \alpha_1 \dots \alpha_s)}\, dq_1 + \int \sqrt{f_2(q_2, \alpha_1 \dots \alpha_s)}\, dq_2 + \dots .$$

The orbit and the time then follow from

$$\frac{\partial S}{\partial \alpha_1} = t + \beta_1, \quad \frac{\partial S}{\partial \alpha_2} = \beta_2, \dots .$$

44. *Illustrative example of motion in a vertical plane under gravity.* If x, y are the horizontal and vertical coordinates,

$$T = \tfrac{1}{2}m(\dot{x}^2 + \dot{y}^2) \text{ and } V = mgy,$$

$$\left.\begin{aligned} p_1 &= \frac{\partial T}{\partial \dot{x}} = m\dot{x} \\ p_2 &= \frac{\partial T}{\partial \dot{y}} = m\dot{y} \end{aligned}\right\},$$

$$\therefore \; H = \frac{1}{2m}(p_1^2 + p_2^2) + mgy.$$

The Hamilton-Jacobi equation is

$$\frac{1}{2m}\left\{\left(\frac{\partial S}{\partial x}\right)^2 + \left(\frac{\partial S}{\partial y}\right)^2\right\} + mgy = \alpha_1,$$

or
$$\left(\frac{\partial S}{\partial x}\right)^2 + \left(\frac{\partial S}{\partial y}\right)^2 + 2m^2gy = 2m\alpha_1.$$

By Hamilton's equations, $\frac{\partial H}{\partial x} = \dot{p}_1$, therefore $\dot{p}_1 = 0$ and p_1 is constant; or, $\frac{\partial S}{\partial x} = \text{constant} = \alpha_2$, say.

$$\therefore \; \left(\frac{\partial S}{\partial y}\right)^2 = 2m\alpha_1 - \alpha_2^2 - 2m^2gy.$$

Thus the variables are separable, and

$$S = \int \frac{\partial S}{\partial x} dx + \int \frac{\partial S}{\partial y} dy,$$

$$S = \alpha_2 x + \int \sqrt{2m\alpha_1 - \alpha_2^2 - 2m^2 gy}.$$

The orbit is given by $\frac{\partial S}{\partial \alpha_2} = \beta_2$ and the time by $\frac{\partial S}{\partial \alpha_1} = t + \beta_1$, i.e. the orbit is

$$x - \int \frac{\alpha_2 dy}{\sqrt{2m\alpha_1 - \alpha_2^2 - 2m^2 gy}} = \beta_2$$

and the time is given by

$$\int \frac{m\,dy}{\sqrt{2m\alpha_1 - \alpha_2^2 - 2m^2 gy}} = t + \beta_1,$$

or $$x + \frac{\alpha_2}{gm^2}\sqrt{2m\alpha_1 - \alpha_2^2 - 2m^2 gy} = \beta_2$$

and $$-\frac{1}{2mg}\sqrt{2m\alpha_1 - \alpha_2^2 - 2m^2 gy} = t + \beta_1.$$

The former gives the parabolic path and the latter the time at which the height is y.

45. *The quantity* $I \equiv \oint pdq$.

In § 43, $\frac{\partial S}{\partial q_k} = \sqrt{f_k(q_k, \alpha_1 \ldots \alpha_s)}$ and $p_k = \frac{\partial S}{\partial q_k}$.

If the α's satisfy the condition that every function $f(q, \alpha_1 \ldots \alpha_s)$ possesses at least two successive real simple roots $q_{max.}$, $q_{min.}$ between which the value of the function is positive, the expression under the square root has the factors $(q_{max.} - q)(q - q_{min.})$, and for p_k to be real, q oscillates between $q_{max.}$ and $q_{min.}$, a maximum and minimum value.

Denote by $\oint pdq$ the value of the integral taken from $q_{min.}$ to $q_{max.}$ and back again to $q_{min.}$, i.e. through a 'libration' of q, and let

$$I_k = \oint p_k dq_k = \oint \sqrt{f_k(q_k, \alpha_1 \ldots \alpha_s)}\,.\,dq_k \quad \ldots\ldots(1).$$

Then $\int p_k dq_k$ increases by I_k during a 'libration' of q_k, and since

$$S = \int p_1 dq_1 + \ldots + \int p_k dq_k + \ldots,$$

it follows that S increases by I_k during a 'libration' of q_k. I_k is called a 'modulus of periodicity' of S.

Again in (1) the integral is taken between certain limits for q_k, so that I_k is a function of the α's. Hence the I's are all functions of the α's and inversely the α's can be found in terms of the I's. Thus in particular α_1 is a function of the I's, i.e. H, the total energy, which is equal to α_1, is also a function of the I's.

It will be shown later that if the system is 'non-degenerate,' that is, the number of frequencies is equal to the number of coordinates, the quantities I are 'adiabatic invariants' and may be equated to whole multiples of h, Planck's constant, so that $I_1 = n_1 h$, $I_2 = n_2 h$, etc.

In this way the total energy ($-W$) can be expressed in terms of s quantum numbers $n_1, n_2 \ldots n_s$, and the nature of the spectrum analysed by Bohr's equation $h\nu = W' - W$.

46. *The Keplerian orbit.* Consider the case of the hydrogen atom for which

$$2T = m(\dot{r}^2 + r^2\dot{\theta}^2), \text{ and } V = -\frac{e^2}{r},$$

where r, θ are the polar coordinates of the electron, the origin being the nucleus.

Also $$p_1 = \frac{\partial T}{\partial \dot{r}} = m\dot{r}, \quad p_2 = \frac{\partial T}{\partial \dot{\theta}} = mr^2\dot{\theta},$$

$$\therefore \; H = T + V = \frac{m}{2}(\dot{r}^2 + r^2\dot{\theta}^2) - \frac{e^2}{r} = \frac{1}{2m}\left(p_1^2 + \frac{p_2^2}{r^2}\right) - \frac{e^2}{r}$$
$$= \alpha_1, \text{ say.}$$

The Hamilton-Jacobi equation is

$$\left(\frac{\partial S}{\partial r}\right)^2 + \frac{1}{r^2}\left(\frac{\partial S}{\partial \theta}\right)^2 - \frac{2me^2}{r} = 2m\alpha_1.$$

The Hamiltonian equation $\frac{\partial H}{\partial \theta} = \dot{p}_2$ gives $\dot{p}_2 = 0$ or $p_2 = \text{constant}$,

$$\therefore \quad \frac{\partial S}{\partial \theta} = \text{constant} = \alpha_2,$$

$$\therefore \quad \left(\frac{\partial S}{\partial r}\right)^2 = 2m\alpha_1 + \frac{2me^2}{r} - \frac{\alpha_2^2}{r^2}.$$

Thus the variables separate.

$$\therefore \quad S = \int \frac{\partial S}{\partial r} dr + \int \frac{\partial S}{\partial \theta} d\theta$$

$$= \alpha_2 \theta + \int dr \sqrt{2m\alpha_1 + \frac{2me^2}{r} - \frac{\alpha_2^2}{r^2}}.$$

The equations $\frac{\partial S}{\partial \alpha_1} = t + \beta_1$ and $\frac{\partial S}{\partial \alpha_2} = \beta_2$ give respectively the time and the form of the orbit (an ellipse).

We now proceed to find I_1 and I_2.

$$I_2 = \oint p_2 d\theta = \oint \alpha_2 d\theta = \int_0^{2\pi} \alpha_2 d\theta = 2\pi\alpha_2,$$

$$I_1 = \oint p_1 dr = \oint dr \sqrt{2m\alpha_1 + \frac{2me^2}{r} - \frac{\alpha_2^2}{r^2}},$$

or

$$I_1 = \oint dr \sqrt{-A + \frac{2B}{r} - \frac{C}{r^2}},$$

where

$$A = -2m\alpha_1, \quad B = me^2, \quad C = \alpha_2^2,$$

$$= \oint \frac{dr}{r} \sqrt{-Ar^2 + 2Br - C}$$

$$= \oint \frac{dr}{r} \sqrt{A(\alpha - r)(r - \beta)},$$

where

$$\alpha\beta = C/A, \quad \alpha + \beta = 2B/A.$$

The α, β correspond to the $q_{\text{max.}}$, $q_{\text{min.}}$ of the general theory of § 45, and r corresponds to q. The symbol $\oint$ means that the integral is taken from β to α and back to β.

$$\therefore \quad I_1 = 2\int_\beta^\alpha \frac{dr}{r} \sqrt{A(\alpha - r)(r - \beta)}.$$

Write $$r = \alpha \sin^2\phi + \beta \cos^2\phi.$$

Then

$$I_1 = 2\int_0^{\frac{\pi}{2}} \frac{\sqrt{A}\,(\alpha-\beta)^2\, 2\sin^2\phi\cos^2\phi}{\alpha\sin^2\phi+\beta\cos^2\phi}\,d\phi$$

$$= 4\sqrt{A}\,(\alpha-\beta)^2 \int_0^{\frac{\pi}{2}} \frac{s^2-s^4}{(\alpha-\beta)\,s^2+\beta}\,d\phi, \text{ where } s = \sin\phi,$$

$$= 4\sqrt{A}\,(\alpha-\beta)^2 \int_0^{\frac{\pi}{2}} \left[\frac{-s^2}{\alpha-\beta} + \frac{\alpha}{(\alpha-\beta)^2} - \frac{\frac{\alpha\beta}{(\alpha-\beta)^2}}{(\alpha-\beta)\,s^2+\beta}\right] d\phi$$

$$= 4\sqrt{A}\,(\alpha-\beta)^2 \left[\frac{-\pi}{4\,(\alpha-\beta)} + \frac{\pi\alpha}{2\,(\alpha-\beta)^2} - \frac{\alpha\beta}{(\alpha-\beta)^2}\int_0^{\frac{\pi}{2}} \frac{d\phi}{\alpha\sin^2\phi+\beta\cos^2\phi}\right]$$

$$= 4\sqrt{A}\,(\alpha-\beta)^2 \left[\frac{-\pi}{4\,(\alpha-\beta)} + \frac{\pi\alpha}{2\,(\alpha-\beta)^2} - \frac{\alpha\beta}{(\alpha-\beta)^2}\,\frac{\pi}{2\sqrt{\alpha\beta}}\right]$$

$$= \pi\sqrt{A}\,(\alpha+\beta-2\sqrt{\alpha\beta})$$

$$= \pi\sqrt{A}\left(\frac{2B}{A} - 2\sqrt{\frac{C}{A}}\right)$$

$$= 2\pi\left(\frac{B}{\sqrt{A}} - \sqrt{C}\right).$$

This result

$$\oint dr\sqrt{-A+\frac{2B}{r}-\frac{C}{r^2}} = 2\pi\left(\frac{B}{\sqrt{A}} - \sqrt{C}\right),$$

will often be required later.

From this, inserting the values of A, B, C,

$$I_1 = 2\pi\left(\frac{me^2}{\sqrt{-2m\alpha_1}} - \alpha_2\right).$$

Also $$I_2 = 2\pi\alpha_2,$$

$$\therefore\ I_1 + I_2 = \frac{2\pi me^2}{\sqrt{-2m\alpha_1}}.$$

But $\alpha_1 = H$, the total energy $= -W$,

$$\therefore \quad W = \frac{2\pi^2 me^4}{(I_1 + I_2)^2},$$

and so as stated in the general theory W can be expressed in terms of the I's only.

47. *Lagrange's method of 'variation of arbitrary constants.'* A problem for which the Hamiltonian function is H_0 is supposed to have been solved for variables p, q, where p is the momentum and q the coordinate, as before.

It is required to solve a problem for which the Hamiltonian function is $H_0 + H_1 \equiv H$. Suppose the solution of the first problem has been found by the use of the Hamilton-Jacobi equation $H_0\left(\frac{\partial S}{\partial q}, q\right) = \alpha_1$, so that the p, q's have been expressed as functions of t, $\alpha_1 \ldots \alpha_s$, $\beta_1 \ldots \beta_s$ from the equations

$$\frac{\partial S}{\partial \alpha_1} = t + \beta_1, \quad \frac{\partial S}{\partial \alpha_2} = \beta_2, \ldots \frac{\partial S}{\partial \alpha_s} = \beta_s.$$

Writing $S' = S - \alpha_1 t$, this Hamilton-Jacobi equation becomes $H_0\left(\frac{\partial S'}{\partial q}, q\right) = -\frac{\partial S'}{\partial t}$, and

$$\frac{\partial S'}{\partial \alpha_1} = \beta_1, \quad \frac{\partial S'}{\partial \alpha_2} = \beta_2, \ldots \frac{\partial S'}{\partial \alpha_s} = \beta_s.$$

Using the contact transformation $\Sigma\,(p\delta q + \beta\delta\alpha) = \delta S'$, we have for the second problem

$$\dot{\alpha} = \frac{\partial K}{\partial \beta}, \quad \dot{\beta} = -\frac{\partial K}{\partial \alpha},$$

where $$K = H + \frac{\partial S'}{\partial t} = H - \alpha_1 = H - H_0 = H_1,$$

$$\therefore \quad \dot{\alpha} = \frac{\partial H_1}{\partial \beta}, \quad \dot{\beta} = -\frac{\partial H_1}{\partial \alpha} \quad \ldots\ldots\ldots\ldots\ldots(1).$$

Thus the effect of a perturbation of the H_0 problem, which adds H_1 to the Hamiltonian function, is determined by the equations (1). From them the values of α, β can be found, and when substituted in the values of q found for the H_0 problem, give the solution of the perturbed $(H_0 + H_1)$ problem.

48. *Infinitesimal contact transformation.* Consider the transformation

$$P = p + \epsilon p', \quad Q = q + \epsilon q' \;..............(1),$$

where p', q' are functions of p, q and ϵ is an infinitesimal constant.

If this is a contact transformation,

$$\Sigma\,(p\delta q - P\delta Q) = \delta F,$$

or
$$\Sigma\,\{p\delta q - (p + \epsilon p')\,(\delta q + \epsilon\delta q')\} = \delta F,$$

or
$$\epsilon\Sigma\,(p'\delta q + p\delta q') = -\,\delta F, \text{ omitting } \epsilon^2.$$

Therefore
$$\Sigma\,\{p'\delta q - q'\delta p + \delta\,(pq')\} = -\,\delta\left(\frac{F}{\epsilon}\right),$$

or
$$\Sigma\,(p'\,\delta q - q'\,\delta p) = -\,\delta\left(\frac{F}{\epsilon} + pq'\right)$$
$$= -\,\delta K, \text{ suppose.}$$

$$\therefore\; p' = -\frac{\partial K}{\partial q} \text{ and } q' = \frac{\partial K}{\partial p}.$$

Thus the general form of infinitesimal contact transformation is given, using (1), by

$$\left.\begin{aligned} P &= p - \epsilon\frac{\partial K}{\partial q} \\ Q &= q + \epsilon\frac{\partial K}{\partial p} \end{aligned}\right\},$$

where K is an arbitrary function of p, q.

If for ϵ we write δt and for P, Q we write $p + \delta p$, $q + \delta q$, then we obtain as a limit, $\dot{p} = -\dfrac{\partial K}{\partial q}$, $\dot{q} = \dfrac{\partial K}{\partial p}$, from which it appears that the canonical equations of motion indicate that the motion of a system in successive small intervals of time corresponds to a succession of infinitesimal contact transformations.

CHAPTER VII

THE USE OF ANGLE VARIABLES; MULTIPLY PERIODIC SYSTEMS

49. *Angle variables w.* In § 45 it was seen that S was a function of the q's and α's and that the α's were functions of the I's, so that $S = f(q, I)$. Also p was equal to $\frac{\partial S}{\partial q}$.

Since $\quad S = f(q, I), \quad \delta S = \Sigma \frac{\partial S}{\partial q} \delta q + \Sigma \frac{\partial S}{\partial I} \delta I$

$$= \Sigma p \delta q + \Sigma \frac{\partial S}{\partial I} \delta I.$$

Now write $w = \frac{\partial S}{\partial I}$. Then $\Sigma (p\delta q + w\delta I) = \delta S$.

This is a contact transformation of the form (β) (§ 41); since S does not contain the time explicitly, the Hamiltonian function is the same for the new variables I, w as for the old variables p, q, and the form of the equations is canonical for I, w, so that

$$\frac{\partial H}{\partial I} = \dot{w}, \quad \frac{\partial H}{\partial w} = -\dot{I}.$$

But $H = \alpha_1$, and is a function of the I's only.

Therefore $\dot{I} = 0$ and I is constant.

Therefore $\frac{\partial H}{\partial I}$, which is also a function of the I's only, is a constant too, equal to ω, say.

Therefore $\dot{w} = \omega$ and $w = \omega t + \epsilon$, where w depends upon the I's only and ϵ is an arbitrary constant.

The quantities w are the 'angle variables' of celestial mechanics.

49 a. *Periodicity of the q's in the w's.* It has been seen that

$$S = f_1(q_1, \alpha) + f_2(q_2, \alpha) + \dots + f_s(q_s, \alpha),$$

where α stands for all the α's.

But the α's are functions of the I's,

$$\therefore\ S = S_1(q_1, I) + S_2(q_2, I) + \ldots + S_s(q_s, I), \quad (1),$$

where I stands for all the I's.

Now $w_k = \dfrac{\partial S}{\partial I_k}$, so that if Δw_k is the change of w_k in a libration of q_m,

$$\begin{aligned}\Delta w_k &= \oint \frac{\partial w_k}{\partial q_m} dq_m \\ &= \oint \frac{\partial^2 S}{\partial I_k \partial q_m} dq_m \\ &= \frac{\partial}{\partial I_k} \oint \frac{\partial S}{\partial q_m} dq_m \\ &= \frac{\partial}{\partial I_k} \oint \frac{\partial S_m}{\partial q_m} dq_m, \text{ from (1)}, \\ &= \frac{\partial}{\partial I_k} \oint p_m dq_m \\ &= \frac{\partial}{\partial I_k}(I_m),\end{aligned}$$

which $= 0$, if $m \neq k$; or $= 1$, if $m = k$.

Therefore w_k increases by unity in a libration of q_k, but by zero in a libration of any other q.

Hence in a libration of q_1, w_1 increases by unity and w_2, $w_3, \ldots$ do not change.

Conversely, if w_1 increases by unity and the other w's do not alter, there is a libration of q_1 only (for if any other q librated, the corresponding w would increase by unity). In this case any q (other than q_1) would change but return to its initial value by retracing its steps without going through a cycle, for it is infinitely improbable that $w_2, w_3, \ldots$ would remain unchanged unless $q_2, q_3, \ldots$ all returned to their original values.

Hence if w_1 increases by unity and the other w's do not alter, all the q's are unchanged, so that each q is periodic in w_1, with period unity; so for each w.

Thus the q's are functions of the w's and the I's periodic in each w with period unity.

50. *The function* $S^* \equiv S - \Sigma Iw$. S increases by I_k in a libration of q_k.

ΣIw also increases by I_k in a libration of q_k, because w_k increases by unity in that case and the other w's do not alter.

Therefore $S - \Sigma Iw$ is unchanged in a libration of any q, or what is the same, when any w increases by unity.

Therefore $S - \Sigma Iw$ is periodic in the w's of period unity.

Or, S^* is a periodic function of the w's, of period unity in each w.

51. *Extension of Fourier's theorem to multiply periodic functions.* It has been seen that the q's are functions of the w's and I's and are periodic with period unity in each of the w's.

The ordinary statement of Fourier's theorem for a single variable θ is

$$f(\theta) = \sum_{\tau=0}^{\infty} (A_\tau \cos \tau\theta + B_\tau \sin \tau\theta),$$

where τ is any positive integer and the range of θ is of extent 2π; but if $f(\theta)$ is periodic in θ of period 2π the theorem holds for any range of θ.

This result may be written $f(\theta) = \sum_{-\infty}^{\infty} C_\tau e^{i\tau\theta}$, as is seen by writing exponential forms for the sine and cosine.

To find C_τ multiply both sides by $e^{-i\tau\theta}$ and integrate from 0 to 2π,

$$\therefore \int_0^{2\pi} f(\theta)\, e^{-i\tau\theta} d\theta = 2\pi C_\tau,$$

which determines C_τ.

Now write $2\pi w = \theta$, so that $f(2\pi w)$ has a period unity in w. Then the above become

$$f(2\pi w) = \sum_{-\infty}^{\infty} C_\tau e^{2\pi i\tau w},$$

where $$2\pi C_\tau = \int_0^1 f(2\pi w)\, e^{-2\pi i\tau w}\, (2\pi dw).$$

Let $f(2\pi w) \equiv F(w)$, so that F is periodic in w with period 1.

Then $$F(w) = \sum_{-\infty}^{\infty} C_\tau e^{2\pi i\tau w},$$

where $$C_\tau = \int_0^1 F(w)\, e^{-2\pi i\tau w} dw.$$

Now consider $F(w, w')$, where F is periodic in each of w, w' of period 1.

Since it is periodic in w', $F(w, w') = \sum_{-\infty}^{\infty} C_{\tau'} e^{2\pi i\tau' w'}$,

where $$C_{\tau'} = \int_0^1 F(w, w')\, e^{-2\pi i\tau' w'} dw',$$

so that $C_{\tau'}$ is a function of w.

$C_{\tau'}$ will be periodic in w with period 1 and is therefore expressible as

$$\sum_{-\infty}^{\infty} D_{\tau,\tau'} e^{2\pi i\tau w}, \text{ where } D_{\tau,\tau'} = \int_0^1 C_{\tau'} e^{-2\pi i\tau w} dw,$$

and is a constant.

$$\therefore\ F(w, w') = \sum_{-\infty}^{\infty} \sum_{-\infty}^{\infty} D_{\tau,\tau'} e^{2\pi i(\tau w + \tau' w')} \quad \ldots\ldots\ldots\ldots\ldots\ldots (1),$$

where $$D_{\tau,\tau'} = \int_0^1 \int_0^1 F(w, w')\, e^{-2\pi i(\tau w + \tau' w')} dw dw'.$$

52. *Multiply periodic system.* In § 49 it was seen that the q's are functions of the I's and the w's, periodic in the w's with period unity. Hence from (1) extended to s variables, each

$$q = \Sigma\Sigma \ldots D_{\tau_1,\tau_2 \ldots \tau_s} e^{2\pi i(\tau_1 w_1 + \tau_2 w_2 + \ldots + \tau_s w_s)},$$

where the D's are functions of the I's.

Since $w_1 = \omega_1 t + \epsilon_1$, $w_2 = \omega_2 t + \epsilon_2$, ..., it follows that

$$q = \Sigma\Sigma \ldots C_{\tau_1 \ldots \tau_s} e^{2\pi i(\tau_1\omega_1 + \tau_2\omega_2 + \ldots + \tau_s\omega_s)t},$$

where the C's are functions of the I's.

Thus the motion of the system may be resolved into a number of harmonic vibrations of frequencies

$$|\,\tau_1\omega_1 + \tau_2\omega_2 + \ldots + \tau_s\omega_s\,|,$$

the amplitudes of which depend on the I's only.

Also the fundamental frequencies $\omega_1, \omega_2, \dots \omega_s$ are given by $\dot{w} = \omega$. But $\dot{w} = \frac{\partial H}{\partial I}$; so that $\omega = \frac{\partial H}{\partial I}$, and the frequencies follow directly from the expression for H in terms of the I's.

53. *The equation* $\delta E = \Sigma \omega \delta I$. Since H is a function of the I's only,

$$\delta H = \Sigma \frac{\partial H}{\partial I} \delta I = \Sigma \omega \delta I,$$

from the preceding article.

And H is the total energy E,

$$\therefore \quad \delta E = \Sigma \omega \delta I.$$

54. *The equation* $\bar{A} = \Sigma \omega I$. The quantity A is defined as $\Sigma p\dot{q}$. In Newtonian mechanics it is equal to $2T$ and is Σmv^2; in relativity mechanics it is (§ 36)

$$\Sigma \frac{\partial T'}{\partial \dot{q}} \dot{q} \text{ or } \Sigma \dot{q} \frac{m_0 c^2}{\sqrt{1-\beta^2}} \beta \frac{\partial \beta}{\partial \dot{q}} \text{ or } \Sigma \frac{m_0 c^2 \beta^2}{\sqrt{1-\beta^2}}, \quad \text{(p. 55)}$$

or

$$m_0 v^2 / \sqrt{1-\beta^2},$$

since

$$\beta = v/c.$$

Now

$$\Sigma p \delta q + \Sigma w \delta I = \delta S,$$

$$\therefore \quad \Sigma p \delta q - \Sigma I \delta w = \delta S - \delta (\Sigma I w)$$

$$= \delta S^* \text{ (§ 50)} \quad \dots\dots\dots\dots(1).$$

If $\bar{A}$ is the mean value of A over a long period of time θ sufficient to include a large number of periods of the w's,

$$\bar{A} = \frac{\int_0^\theta \Sigma p\dot{q} dt}{\theta} = \frac{\int_0^\theta \left(\Sigma I \dot{w} + \frac{dS^*}{dt} \right) dt}{\theta}, \text{ from (1)},$$

$$\therefore \quad \bar{A} = \frac{1}{\theta} \int_0^\theta \left(\Sigma I \omega + \frac{dS^*}{dt} \right) dt$$

$$= \Sigma I \omega + \frac{1}{\theta} \Big[S^* \Big]_0^\theta .$$

Since S^* is periodic in the w's (§ 50) and the interval θ contains a large number of periods of the w's, S^* has the same value at both limits, so that

$$\bar{A} = \Sigma I \omega.$$

This and the preceding result play an important part in Bohr's recent exposition of his theory[1].

55. *Degenerate systems.* The frequencies of a system are given by $\omega = \frac{\partial H}{\partial I}$, and in general there are as many independent frequencies ω as there are coordinates.

If however there is a relation between the ω's of the form $\kappa_1\omega_1 + \kappa_2\omega_2 + \ldots + \kappa_s\omega_s = 0$, where the κ's are integers, there are only $(s-1)$ independent frequencies and the system is said to be 'simply' degenerate.

In this case

$$\kappa_1\frac{\partial H}{\partial I_1} + \kappa_2\frac{\partial H}{\partial I_2} + \ldots = 0, \text{ using } \omega = \frac{\partial H}{\partial I} \quad \ldots\ldots(1).$$

This partial differential equation is solved by the auxiliary equations

$$dI_1/\kappa_1 = dI_2/\kappa_2 = \ldots = dI_s/\kappa_s = dH/0,$$

of which independent integrals are

$$I_1/\kappa_1 - I_s/\kappa_s = \text{const.};\ I_2/\kappa_2 - I_s/\kappa_s = \text{const.}; \ldots$$
$$I_{s-1}/\kappa_{s-1} - I_s/\kappa_s = \text{const.};\quad H = \text{const.}$$

Therefore the general integral of (1) is

$$H = \Phi\,[I_1/\kappa_1 - I_s/\kappa_s,\ \ I_2/\kappa_2 - I_s/\kappa_s, \ldots,\ \ I_{s-1}/\kappa_{s-1} - I_s/\kappa_s].$$

Writing $I_1/\kappa_1 - I_s/\kappa_s = J_1$, $I_2/\kappa_2 - I_s/\kappa_s = J_2$, and so on,

$$H = \Phi\,(J_1, J_2, \ldots J_{s-1}).$$

The new angle variables v are connected with the old ones w by

$$w_1 = \frac{\partial S}{\partial I_1} = \frac{\partial S}{\partial J_1}\frac{\partial J_1}{\partial I_1} + \frac{\partial S}{\partial J_2}\frac{\partial J_2}{\partial I_1} + \ldots + \frac{\partial S}{\partial J_{s-1}}\frac{\partial J_{s-1}}{\partial I_1}$$
$$= v_1\frac{1}{\kappa_1} + v_2 . 0 + \ldots + v_{s-1} . 0,$$

or $\quad w_1 = v_1/\kappa_1.$

So $\quad w_2 = v_2/\kappa_2, \ldots$ and $w_{s-1} = v_{s-1}/\kappa_{s-1},$

[1] N. Bohr, 'The effect of electric and magnetic fields on spectral lines.' The Guthrie Lecture to the Physical Society of London, 1922, published in 1923.

but $$w_s = \frac{\partial S}{\partial J_1}\frac{\partial J_1}{\partial I_s} + \text{etc.}$$

$$= v_1\left(-\frac{1}{\kappa_s}\right) + v_2\left(-\frac{1}{\kappa_s}\right) + \ldots + v_{s-1}\left(-\frac{1}{\kappa_s}\right),$$

Hence $w_1 = v_1/\kappa_1,\ w_2 = v_2/\kappa_2, \ldots\ w_{s-1} = v_{s-1}/\kappa_{s-1}$,

and $$w_s = -(v_1 + v_2 + \ldots + v_{s-1})/\kappa_s.$$

Thus the system is by this linear transformation reduced to $(s-1)$ pairs of variables J, v, equal in number to the number of independent frequencies. The quantum conditions are $J = nh$ for each J, so that the number of conditions is equal to the number of independent frequencies.

Again if s is the number of coordinates q, and u the number of independent frequencies, there must be $(s-u)$ relations of the type $\kappa_1\omega_1 + \kappa_2\omega_2 + \ldots + \kappa_s\omega_s = 0$ which can be successively used, as above, to reduce the number of angle variables from s to u, so that $w_1 \ldots w_s$ are linear in $v_1 \ldots v_u$.

Since $$q = \Sigma\Sigma C_{\tau_1\ldots\tau_s} e^{2\pi i(\tau_1 w_1 + \ldots + \tau_s w_s)},$$

it follows that this transforms into

$$\Sigma\Sigma D_{\tau_1'\ldots\tau_s'} e^{2\pi i(\tau_1' v_1 + \ldots + \tau_u' v_u)},$$

or $$\Sigma\Sigma E_{\tau_1'\ldots\tau_s'} e^{2\pi i(\tau_1'\omega_1' + \ldots + \tau_u'\omega_u')t}.$$

There will be u new J's corresponding one to each new v; by writing each J equal to nh, the stationary states are found; there are thus u quantum numbers $n_1, n_2, \ldots n_u$ each corresponding to an independent frequency.

56. An illustration of a degenerate case is that of the Keplerian ellipse (§ 46), where W was found equal to

$$\frac{2\pi^2 me^4}{(I_1 + I_2)^2}.$$

Here $$H = -W = -2\pi^2 me^4/(I_1 + I_2)^2.$$

The frequencies are

$$\omega_1 = \partial H/\partial I_1,\quad \omega_2 = \partial H/\partial I_2,$$

so that $$\omega_1 = \omega_2 = \frac{4\pi^2 me^4}{(I_1 + I_2)^3}.$$

Writing
$$\left.\begin{array}{r} I_1 + I_2 = J_1 \\ I_2 = J_2 \end{array}\right\}$$

and supposing v_1, v_2 to be the new angle variables corresponding to J_1, J_2, we have

$$w_1 = \frac{\partial S}{\partial I_1} = \frac{\partial S}{\partial J_1}\frac{\partial J_1}{\partial I_1} + \frac{\partial S}{\partial J_2}\frac{\partial J_2}{\partial I_1} = v_1,$$

$$w_2 = \frac{\partial S}{\partial I_2} = \frac{\partial S}{\partial J_1}\frac{\partial J_1}{\partial I_2} + \frac{\partial S}{\partial J_2}\frac{\partial J_2}{\partial I_2} = v_1 + v_2,$$

$$\therefore\ v_1 = w_1, \quad v_2 = w_2 - w_1.$$

Since $w_2 - w_1 = (\omega_2 t + \epsilon_2) - (\omega_1 t + \epsilon_1)$ and $\omega_1 = \omega_2$, it follows that $w_2 - w_1 = \epsilon_2 - \epsilon_1$ and is constant.

Thus
$$\left.\begin{array}{l} v_1 = \omega_1 t + \epsilon_1 \\ v_2 = \text{constant} \end{array}\right\}.$$

Also $w = \dfrac{2\pi^2 me^4}{J_1^{\,2}}$, and v_1 is the single angle variable corresponding to J_1. The quantum condition is $J_1 = nh$, from which $w = \dfrac{2\pi^2 me^4}{n^2h^2}$.

57. *The correspondence principle for a multiply periodic system.* The coordinates q of the system can be expressed (§ 55) in the form

$$\Sigma\Sigma C_{\tau_1 \ldots \tau_{s_u}} e^{2\pi i(\tau_1\omega_1 + \ldots + \tau_u\omega_u)},$$

where $\omega_1 \ldots \omega_u$ are the independent frequencies.

Thus the coordinates x, y, z of any particle of the system can be expressed in the same form, as they are functions of the q's.

The components of the electric moment of the system are $\Sigma e\dot{x}$, etc., and have therefore too the same form. It is the variation of the electric moment with the time which in the classical theory determines the constitution of the emitted radiation.

Therefore the frequencies on the classical theory are

$$|\, \tau_1\omega_1 + \ldots + \tau_u\omega_u \,| \equiv \Omega,$$

where the τ's are integers ranging from $-\infty$ to ∞, and are not all zero simultaneously.

If $J_1, J_2, \ldots J_u$ correspond to the 'reduced' angle variables $v_1 \ldots v_u$ of § 55, the quantum conditions are

$$J_1 = n_1 h, \quad J_2 = n_2 h, \ldots.$$

On Bohr's theory, for a transition from a state $(n_1, n_2, \ldots n_u)$ to a state $(n_1', n_2', \ldots n_u')$,

$$\nu = \frac{1}{h}(E - E') = \frac{1}{h}\sum_{s'}^{s} \delta E,$$

where the symbol s denotes the first state and s' the second one.

But $$\delta E = \omega_1 \delta J_1 + \ldots + \omega_u \delta J_u,$$

$$\therefore \quad \nu = \frac{1}{h}\sum_{s'}^{s} (\omega_1 \delta J_1 + \ldots + \omega_u \delta J_u).$$

If each quantum number is large compared with its difference due to the transition, the motions in the two states differ little from one another, so that the ω's are approximately constant in the summation, and

$$\nu = \frac{1}{h}\{\omega_1 \sum_{s'}^{s} \delta J_1 + \ldots\}$$

$$= \frac{1}{h}\{\omega_1 (J_1 - J_1') + \ldots\}$$

$$= \omega_1 (n_1 - n_1') + \omega_2 (n_2 - n_2') + \ldots + \omega_u (n_u - n_u').$$

Thus for large quantum numbers the frequencies are the same as on the classical theory, τ_1 being $n_1 - n_1'$, etc.

In § 23 the significance of this was fully gone into, and if a coefficient τ_1 is zero, i.e. the corresponding ω_1 is absent in the classical frequency, then $n_1 - n_1'$ is zero and the transition in question cannot occur. If, for example, $|\tau_1| = 1$, then $|n_1 - n_1'|$ can only be 1, and n_1 can only change by ± 1.

Thus the Fourier series for the coordinates (or what is effectively the same, the electric moment) indicates the nature of the transitions or 'switches' possible on the quantum theory. This use of the principle will be amply illustrated in later chapters.

58. *Burgers' proof of the invariance of the I's for a non-degenerate system*[1]. Suppose the system to contain a parameter a (such as the length of the thread of the pendulum in § 29) which varies slowly and erratically with the time.

The contact transformation

$$\Sigma\,(p\delta q + w\delta I) = \delta S \qquad \text{..............(1)}$$

is also $$\Sigma\,(p\delta q - I\delta w) = \delta\,(S - \Sigma I w) = \delta S^* \qquad \text{......(2)}$$

and therefore

$$\frac{\partial K}{\partial I} = \dot{w}, \quad \frac{\partial K}{\partial w} = -I,$$

where $$K = H + \frac{\partial S^*}{\partial t}.$$

Since S^* is a function of q, w, from the form of (2), it is a function of a, and through the a varies with the time, so that K is not equal to H.

Since $H = \alpha_1$ = a function of the I's,

$$\frac{\partial H}{\partial w} = 0 \text{ ....(3)},$$

$$\therefore\ \dot{I} = -\frac{\partial K}{\partial w} = -\frac{\partial H}{\partial w} - \frac{\partial^2 S^*}{\partial w \partial t} = -\frac{\partial^2 S^*}{\partial w \partial t}, \text{ from (3).}$$

Also $\dfrac{\partial S^*}{\partial t} = \dfrac{\partial S^*}{\partial a}\,\dot{a} = \dot{a}\phi$, where ϕ is written for $\dfrac{\partial S^*}{\partial a}$.

$$\therefore\ \dot{I} = -\frac{\partial}{\partial w}(\dot{a}\phi) = -\dot{a}\frac{\partial \phi}{\partial w}.$$

Integrating with respect to the time from $t = t_1$ to $t = t_2$, a long interval of time,

$$I_2 - I_1 = -\int_{t_1}^{t_2} \dot{a}\,\frac{\partial \phi}{\partial w}\,dt.$$

But S^* is periodic in each of the w's with period 1 (§ 50).

[1] J. M. Burgers, Versl. Akad. Amsterdam, **25**, p. 1055, 1917; Phil. Mag. **33**, p. 514, 1917. Also N. Bohr, 'Q.L.S.' Part I, pp. 21, 22.

Therefore $\frac{\partial S^*}{\partial a}$, or ϕ, is too.

Therefore ϕ may be written as

$$\Sigma C e^{2\pi i (\tau_1 w_1 + \ldots + \tau_u w_u)}.$$

$$\therefore \quad \frac{\partial \phi}{\partial w} = \Sigma' D e^{2\pi i (\tau_1 w_1 + \ldots + \tau_u w_u)},$$

the dash denoting the absence of the constant term, which the differentiation has removed.

$$\therefore \quad \frac{\partial \phi}{\partial w} = \Sigma' E e^{2\pi i (\tau_1 \omega_1 + \ldots + \tau_u \omega_u) t},$$

and $$I_2 - I_1 = -\int_{t_1}^{t_2} \dot{a} \Sigma' E e^{2\pi i (\tau\omega) t}\, dt,$$

where the symbol $(\tau\omega)$ denotes $\tau_1 \omega_1 + \ldots + \tau_u \omega_u$.

$$\therefore \quad I_2 - I_1 = -(\dot{a})_m \int_{t_1}^{t_2} \Sigma' E e^{2\pi i (\tau\omega) t}\, dt,$$

where $(\dot{a})_m$ is the mean value of $\dot{a}$ in the interval t_1 to t_2.

The expression $\int_{t_1}^{t_2} \Sigma' E e^{2\pi i (\tau\omega) t}\, dt$ contains a through the E's and the ω's and may be written $\int_{t_1}^{t_2} F(a, t)\, dt$, which is equal to $\int_{t_1}^{t_2} [F(a_1, t) + (a - a_1) F'(a_1, t) + \ldots]\, dt$, where a_1 is the value of a at time t_1.

The first term of the integrand is periodic in the constant ω's which existed before a began to vary, and if $t_2 - t_1$ is long enough to include a large number of these periods,

$$\int_{t_1}^{t_2} F(a_1, t)\, dt = 0,$$

on account of the periodic character of F.

The second term

$$\int_{t_1}^{t_2} (a - a_1) F'(a_1, t)\, dt$$

is of the same order as

$$\int_{t_1}^{t_2} (\dot{a} t) F'(a_1, t)\, dt,$$

and is of order $(\dot{a})_m (t_2 - t_1)$, i.e. of order α, where α is the finite change of a in the interval.

Hence $\quad I_2 - I_1 = -(\dot{a})_m$ {term of order α}.

For an infinitely slow variation of a, $(\dot{a})_m \to 0$ and $I_2 = I_1$. Thus the adiabatic invariance of I is proved, in general.

But if there is a relation $\kappa_1\omega_1 + \ldots + \kappa_u\omega_u = 0$ between the original frequencies, where the κ's are integers, then $F(a_1, t)$, which is $\Sigma' \{Ee^{2\pi i(\tau\omega)t}\}_{a=a_1}$, contains a constant term which occurs when $\tau_1 = \kappa_1, \ldots \tau_1 = \kappa_n$, and if this term is C, the integral $\int_{t_1}^{t_2} F(a_1, t)\, dt$ is no longer zero but is equal to $C(t_2 - t_1)$.

$$
\begin{aligned}
\text{Hence}\quad I_2 - I_1 &= -(\dot{a})_m \{C(t_2 - t_1) + \text{a term of order } \alpha\} \\
&= -C\alpha - (\dot{a})_m (\text{term of order } \alpha) \\
&= -C\alpha, \text{ as } (\dot{a})_m \to 0.
\end{aligned}
$$

Therefore I undergoes a finite change in the interval and is not invariant. This is the case of a 'degenerate' system.

Hence I is invariant only for systems so reduced that the number of I's and w's is equal to the number of independent frequencies.

The invariance of the I's is dependent upon the following conditions: (i) the q's are periodic in all the w's of period unity, (ii) the total energy H is a function of the I's only, (iii) S^* is periodic in all the w's of period unity.

59. *The quantum conditions for multiply periodic systems.* If the system is reduced to dependence on $I_1 \ldots I_u, w_1 \ldots w_u$, where u is the number of independent frequencies $\omega_1 \ldots \omega_u$, the quantum conditions are $I_1 = n_1 h, \ldots I_u = n_u h$, for the I's are adiabatic invariants. These conditions were used by Sommerfeld and Epstein in their work on multiply periodic systems, but they used them sometimes for degenerate systems; using s coordinates, they found s I's and equated each of these to an nh, thus imposing s quantum conditions. But if the number of frequencies u is less than s (degenerate system), it is possible, as has been seen, by a linear transformation to find new I's and w's equal in number to the independent frequencies.

It is these new I's, u in number, which should be equated to nh's to form the quantum conditions, for it is only the new I's which are adiabatic invariants, their number being equal to the number of frequencies.

It is apparent that the imposition of s conditions unduly restricts the number of stationary states, which depend upon only u conditions.

CHAPTER VIII

THE RELATIVITY THEORY OF THE FINE STRUCTURE OF THE HYDROGEN LINES; ELECTRON IN A CENTRAL FIELD

60. *The fine structure of the hydrogen lines.* If the lines of the hydrogen spectrum are examined by instruments of high dispersive power, it is found that each line consists of three lines, which are seen as one with ordinary power. Sommerfeld[1] explained this 'fine structure' of the lines by taking into account the change of mass with velocity as required by the theory of relativity.

Sommerfeld's theory[2]. Using plane polar coordinates r, θ,

$$H = T + V = m_0c^2\left[\frac{1}{\sqrt{1-\beta^2}} - 1\right] - \frac{e^2}{r},$$

where $$\beta^2 = \frac{v^2}{c^2} = \frac{\dot{r}^2 + r^2\dot{\theta}^2}{c^2} \quad (\S\ 36).$$

The function

$$T' = m_0c^2[1-\sqrt{1-\beta^2}] = m_0c^2\left[1 - \sqrt{1 - \frac{\dot{r}^2 + r^2\dot{\theta}^2}{c^2}}\right],$$

$$\therefore\ p_1 = \frac{\partial T'}{\partial \dot{r}} = \frac{m_0\dot{r}}{\sqrt{1-\beta^2}} \text{ and } p_2 = \frac{\partial T'}{\partial \dot{\theta}} = \frac{m_0 r^2\dot{\theta}}{\sqrt{1-\beta^2}},$$

$$\therefore\ p_1{}^2 + \frac{p_2{}^2}{r^2} = \frac{m_0{}^2\beta^2c^2}{1-\beta^2},$$

$$\therefore\ m_0{}^2c^2 + p_1{}^2 + \frac{p_2{}^2}{r^2} = \frac{m_0{}^2c^2}{1-\beta^2},$$

$$\therefore\ H = c\sqrt{\left(m_0{}^2c^2 + p_1{}^2 + \frac{p_2{}^2}{r^2}\right)} - m_0c^2 - \frac{e^2}{r} = \alpha_1,$$

where α_1 is the total energy.

[1] A. Sommerfeld, Berich. Akad. München, pp. 425, 459, 1915; p. 131, 1916; p. 83, 1917; Ann. d. Phys. 51, p. 1, 1916.

[2] A. Sommerfeld, 'Atomic Structure and Spectral Lines,' chap. VIII (English translation, 1923).

$$\therefore\ c^2\left(m_0{}^2c^2+p_1{}^2+\frac{p_2{}^2}{r^2}\right)=\left(\alpha_1+\frac{e^2}{r}+m_0c^2\right)^2,$$

$$c^2\left(p_1{}^2+\frac{p_2{}^2}{r^2}\right)=\left(\alpha_1+\frac{e^2}{r}\right)^2+2m_0c^2\left(\alpha_1+\frac{e^2}{r}\right),$$

$$c^2\left(p_1{}^2+\frac{p_2{}^2}{r^2}\right)=\left(\alpha_1+\frac{e^2}{r}\right)\left(\alpha_1+\frac{e^2}{r}+2m_0c^2\right).$$

Writing $$p_1=\frac{\partial S}{\partial r},\quad p_2=\frac{\partial S}{\partial\theta},$$

we have the Hamilton-Jacobi equation

$$\left(\frac{\partial S}{\partial r}\right)^2+\frac{1}{r^2}\left(\frac{\partial S}{\partial\theta}\right)^2=\frac{1}{c^2}\left(\alpha_1+\frac{e^2}{r}\right)\left(\alpha_1+\frac{e^2}{r}+2m_0c^2\right).$$

Since $\dfrac{\partial H}{\partial\theta}=-\dot{p}_2$ and H does not contain θ, $\dot{p}_2=0$ or p_2 is constant and equal to α_2.

$$\therefore\ \frac{\partial S}{\partial\theta}=\alpha_2.$$

Therefore

$$\left(\frac{\partial S}{\partial r}\right)^2=-\frac{\alpha_2{}^2}{r^2}+\frac{1}{c^2}\left(\alpha_1+\frac{e^2}{r}\right)\left(\alpha_1+\frac{e^2}{r}+2m_0c^2\right)$$

$$=\left(2m_0+\frac{\alpha_1}{c^2}\right)\alpha_1+\frac{2e^2}{r}\left(m_0+\frac{\alpha_1}{c^2}\right)-\frac{1}{r^2}\left(\alpha_2{}^2-\frac{e^4}{c^2}\right)$$

$$=-A+\frac{2B}{r}-\frac{C}{r^2},$$

where

$$A=-\left(2m_0+\frac{\alpha_1}{c^2}\right)\alpha_1,\quad B=e^2\left(m_0+\frac{\alpha_1}{c^2}\right),\quad C=\alpha_2{}^2-\frac{e^4}{c^2}.$$

Now

$$I_1=\oint p_1dr=\oint dr\sqrt{-A+\frac{2B}{r}-\frac{C}{r^2}}=2\pi\left(\frac{B}{\sqrt{A}}-\sqrt{C}\right)\ (\S\ 46),$$

and $$I_2=\oint p_2d\theta=\int_0^{2\pi}\alpha_2d\theta=2\pi\alpha_2.$$

Writing in the values of A, B, C,

$$I_1=\frac{2\pi e^2\left(m_0+\dfrac{\alpha_1}{c^2}\right)}{\sqrt{-\alpha_1\left(2m_0+\dfrac{\alpha_1}{c^2}\right)}}-2\pi\sqrt{\alpha_2{}^2-\frac{e^4}{c^2}}.$$

Denote the magnitude $2\pi e^2/ch$ by μ; using the known values of e, c, and h, $\mu^2 = 5{\cdot}31 \times 10^{-5}$. (If e is in E.S.U. μ is a number without dimensions in mass, length, time.)

Then $$I_1 + \sqrt{I_2^2 - h^2\mu^2} = \frac{ch\mu\left(m_0 + \frac{\alpha_1}{c^2}\right)}{\sqrt{-\alpha_1\left(2m_0 + \frac{\alpha_1}{c^2}\right)}}$$

$$= \frac{h\mu\left(m_0 + \frac{\alpha_1}{c^2}\right)}{\sqrt{m_0^2 - \left(m_0 + \frac{\alpha_1}{c^2}\right)^2}}.$$

Write $$1 + \frac{\alpha_1}{m_0 c^2} = U.$$

Then $$I_1 + \sqrt{I_2^2 - h^2\mu^2} = \frac{h\mu U}{\sqrt{1 - U^2}},$$

$$\therefore \quad \frac{1 - U^2}{U^2} = \frac{h^2\mu^2}{(I_1 + \sqrt{I_2^2 - h^2\mu^2})^2},$$

or $$\frac{1}{U^2} = 1 + \frac{h^2\mu^2}{(I_1 + \sqrt{I_2^2 - h^2\mu^2})^2}$$

and $\alpha_1 = -W$, where W is the negative energy, so that

$$\frac{1}{\left(1 - \frac{W}{m_0 c^2}\right)^2} = 1 + \frac{h^2\mu^2}{(I_1 + \sqrt{I_2^2 - h^2\mu^2})^2},$$

which gives W in terms of I_1, I_2.

Approximate to terms containing $\frac{1}{c^4}$, noting that μ^2 contains $\frac{1}{c^2}$. It follows that

$$\left(1 - \frac{W}{m_0 c^2}\right)^{-2} = 1 + h^2\mu^2 \Big/ \left\{I_1 + I_2\left(1 - \frac{h^2\mu^2}{2I_2^2}\right)\right\}^2, \text{ to this order,}$$

$$= 1 + \frac{h^2\mu^2}{(I_1 + I_2)^2}\left\{1 + \frac{h^2\mu^2}{I_2(I_1 + I_2)}\right\},$$

$$\therefore\ 1-\frac{W}{m_0c^2}=\left[1+\frac{h^2\mu^2}{(I_1+I_2)^2}+\frac{h^4\mu^4}{I_2(I_1+I_2)^3}\right]^{-\frac{1}{2}}$$

$$=1-\frac{h^2\mu^2}{2(I_1+I_2)^2}-\frac{h^4\mu^4}{2I_2(I_1+I_2)^3}+\frac{3}{8}\frac{h^4\mu^4}{(I_1+I_2)^4}$$

$$=1-\frac{h^2\mu^2}{2(I_1+I_2)^2}-\frac{h^4\mu^4(I_2+4I_1)}{8I_2(I_1+I_2)^4},$$

$$\therefore\ W=\frac{m_0c^2h^2\mu^2}{2(I_1+I_2)^2}\left\{1+\frac{h^2\mu^2}{(I_1+I_2)^2}\left(\frac{1}{4}+\frac{I_1}{I_2}\right)\right\}\ldots\ldots(1).$$

The frequencies are $\frac{\partial H}{\partial I_1}, \frac{\partial H}{\partial I_2}$, and since $W=-H$, they are evidently unequal; the system is not degenerate, so that the quantum conditions are $I_1=n_1h$, $I_2=n_2h$.

$$\therefore\ W=\frac{m_0c^2\mu^2}{2(n_1+n_2)^2}\left\{1+\frac{\mu^2}{(n_1+n_2)^2}\left(\frac{1}{4}+\frac{n_1}{n_2}\right)\right\},$$

or writing $n_1+n_2=n$, $n_2=k$, and $\mu^2=4\pi^2e^4/c^2h^2$,

$$W=\frac{2\pi^2m_0e^4}{h^2n^2}\left\{1+\frac{\mu^2}{n^2}\left(\frac{n}{k}-\frac{3}{4}\right)\right\}\ \ldots\ldots\ldots\ldots(2).$$

Thus the energy value of an n_k orbit is found.

61. *Use of the correspondence principle.*

$$S=\int\frac{\partial S}{\partial r}dr+\int\frac{\partial S}{\partial\theta}d\theta$$

$$=\int dr\sqrt{-A+\frac{2B}{r}+\frac{C}{r^2}}+\alpha_2\theta,$$

A, B, C are functions of α_1, α_2 and therefore of I_1, I_2. [Equation (1) gives α_1, which is $-W$, in terms of I_1 and I_2, and $\alpha_2=I_2/2\pi$.]

Therefore the first term on the right-hand side is a function of r, I_1, I_2.

Hence $$S=f(r, I_1, I_2)+\frac{\theta I_2}{2\pi}.$$

Therefore the angle variables w_1, w_2 are given by

$$w_1=\frac{\partial S}{\partial I_1}=F(r, I_1, I_2),$$

$$w_2=\frac{\partial S}{\partial I_2}=\phi(r, I_1, I_2)+\frac{\theta}{2\pi}.$$

$$\therefore\ r = \chi(w_1, I_1, I_2),$$
$$\theta = 2\pi w_2 - \phi(r, I_1, I_2)$$
$$= 2\pi w_2 + \psi(w_1, I_1, I_2),$$

substituting for r in terms of w_1, I_1, I_2.

$$\therefore\ x + iy = re^{i\theta} = \chi(w_1, I_1, I_2)\, e^{2\pi i w_2 + i\psi(w_1, I_1, I_2)}$$
$$= \Phi(w_1, I_1, I_2)\, e^{2\pi i w_2}.$$

Writing $\Phi(w_1, I_1, I_2)$ as a Fourier series $\Sigma C_\tau e^{2\pi i \tau w_1}$, where C_τ is a function of I_1, I_2, we have

$$x + iy = \Sigma C_\tau e^{2\pi i(\tau w_1 + w_2)}$$
$$= \Sigma D_\tau e^{2\pi i(\tau\omega_1 + \omega_2)t},\ \text{since}\ \left.\begin{aligned} w_1 &= \omega_1 t + \epsilon_1 \\ w_2 &= \omega_2 t + \epsilon_2 \end{aligned}\right\}.$$

Thus the coordinates x, y (and therefore the electric moment $\Sigma e\dot{x}$, $\Sigma e\dot{y}$) contain terms of the type $D_\tau e^{2\pi i(\tau\omega_1 + \omega_2)t}$, so that by the correspondence principle, transitions are only possible where $|\Delta n_2|$ is 1, while $|\Delta n_1|$ may have any value. [The coefficient of ω_2 is unity, while that of ω_1 is τ, where τ is unrestricted.]

Thus while n_1 may change by any amount, n_2 can only change by unity.

Writing $n_1 + n_2 = n$, $n_2 = k$ as above, we see that k can only change by unity in a switch, and $k \leqslant n$. When the form of the orbit is worked out in § 64, it will be seen that $k = 0$ gives a path which would pass through the nucleus, and it is presumed that a state in which the electron collides with the nucleus is not a stationary state.

62. *Calculation of the fine structure for hydrogen.*

$$W = \frac{2\pi^2 m_0 e^4}{h^2 n^2}\left\{1 + \frac{\mu^2}{n^2}\left(\frac{n}{k} - \frac{3}{4}\right)\right\},$$

and

$$R = \frac{2\pi^2 m_0 e^4}{ch^3},$$

$$\therefore\ W = Rhc\left\{\frac{1}{n^2} + \frac{\mu^2}{n^4}\left(\frac{n}{k} - \frac{3}{4}\right)\right\},$$

where

$$R = 109678,$$
$$\mu^2 = 5{\cdot}31 \times 10^{-5}.$$

For a switch from an n_k orbit to an $n'_{k'}$ orbit,

$$\nu = \frac{1}{h}(E - E') = \frac{1}{h}(W' - W),$$

$$\therefore \quad \nu = Rc\left\{\frac{1}{n'^2} - \frac{1}{n^2} + \frac{\mu^2}{n'^4}\left(\frac{n'}{k'} - \frac{3}{4}\right) - \frac{\mu^2}{n^4}\left(\frac{n}{k} - \frac{3}{4}\right)\right\}.$$

If ν_0 is the frequency of the line given by the switch $n \rightarrow n'$ if relativity is neglected,

$$\nu_0 = Rc\left(\frac{1}{n'^2} - \frac{1}{n^2}\right),$$

$$\therefore \quad \nu - \nu_0 = Rc\mu^2\left\{\frac{1}{n'^4}\left(\frac{n'}{k'} - \frac{3}{4}\right) - \frac{1}{n^4}\left(\frac{n}{k} - \frac{3}{4}\right)\right\} \quad \text{...(3).}$$

Thus the single line of frequency ν_0 now becomes several lines of frequencies given by (3).

Consider the H_α line of the Balmer series corresponding to $n = 3 \rightarrow n' = 2$. Possible values of k are 1, 2, 3 and of k' 1, 2. But by the correspondence principle $k' - k = \pm 1$ and we have seen that $k \neq 0$ and $k' \neq 0$; therefore only the following pairs of values of k, k' are possible:

$$\left.\begin{matrix} k = 1 \\ k' = 2 \end{matrix}\right\} \quad \left.\begin{matrix} k = 2 \\ k' = 1 \end{matrix}\right\} \quad \left.\begin{matrix} k = 3 \\ k' = 2 \end{matrix}\right\}.$$

These correspond to the three values:

$$\frac{\nu - \nu_0}{Rc\mu^2} = \frac{1}{16}\left(\frac{2}{2} - \frac{3}{4}\right) - \frac{1}{81}\left(\frac{3}{1} - \frac{3}{4}\right) = \frac{-7}{576} \quad (3_1 \text{ orbit} \rightarrow 2_2 \text{ orbit}),$$

or $$\frac{1}{16}\left(\frac{2}{1} - \frac{3}{4}\right) - \frac{1}{81}\left(\frac{3}{2} - \frac{3}{4}\right) = \frac{119}{1728} \quad (3_2 \rightarrow 2_1),$$

or $$\frac{1}{16}\left(\frac{2}{2} - \frac{3}{4}\right) - \frac{1}{81}\left(\frac{3}{3} - \frac{3}{4}\right) = \frac{65}{5184} \quad (3_3 \rightarrow 2_2).$$

To express this in wave length, $\nu = \dfrac{c}{\lambda}$,

$$\therefore \quad \nu - \nu_0 = c\left(\frac{1}{\lambda} - \frac{1}{\lambda_0}\right) = \frac{c(\lambda_0 - \lambda)}{\lambda_0^2}, \text{ approx.}$$

$$\therefore \quad \frac{\lambda - \lambda_0}{R\mu^2\lambda_0^2} = \frac{7}{576} \text{ or } \frac{-119}{1728} \text{ or } \frac{-65}{5184}.$$

But $$\frac{1}{\lambda_0} = R\left(\frac{1}{2^2} - \frac{1}{3^2}\right) = \frac{5R}{36} = \frac{5 \times 109678}{36} \quad ......(\alpha),$$

$$\therefore \lambda_0 = \frac{36}{5 \times 109678} \times 10^8 \text{ Ångström units}$$
$$= 6564 \text{ Å}.$$

$$\therefore R\mu^2\lambda_0^2 \text{ in cm.} = (R\lambda_0)\,\mu^2\lambda_0 = \left(\frac{36}{5}\right)\mu^2\lambda_0 \text{ from } (\alpha)$$
$$= \frac{36}{5}(5{\cdot}31 \times 10^{-5})(6564 \times 10^{-8})$$
$$= \frac{36 \times 5{\cdot}31 \times 6564}{5} \times 10^{-5} \text{ Å}.$$
$$= 2{\cdot}51 \text{ Å}.$$

$$\therefore \lambda - \lambda_0 = \frac{7}{576} \text{ or } \frac{-119}{1728} \text{ or } \frac{-65}{5184} \text{ of } 2{\cdot}51 \text{ Å}.$$
$$= {\cdot}030 \text{ Å. or } -{\cdot}173 \text{ Å. or } -{\cdot}031 \text{ Å.} \quad ...(4).$$

The theoretical values of $\lambda - \lambda_0$ are in the ratio

$$\frac{5}{576} : \frac{-119}{1728} : \frac{-65}{5184} \text{ or } 63 : -357 : -65.$$

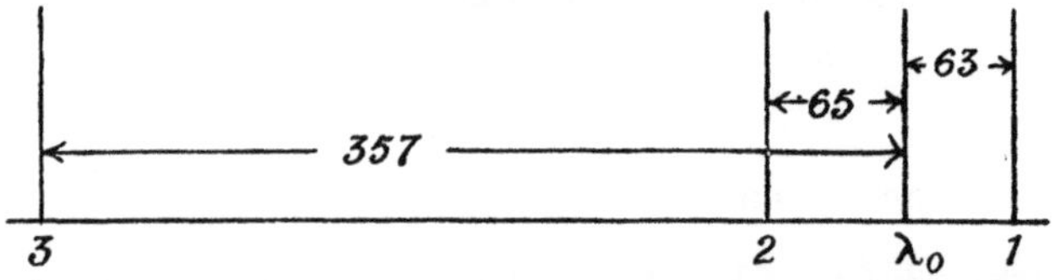

The figure shows the separation and that the separation of 1 and 2 is to that of 2 and 3 in the ratio 292 : 128 or 2·3 : 1.

The theory gives the separation 12 as ·061 Å., from (4).

Now the width of a spectral line due to the Döppler effect can be calculated[1] in terms of the absolute temperature t and the molecular weight M of the gas, and is $\cdot 86 \times 10^{-6}\,\lambda\sqrt{t/M}$; and for hydrogen with its low molecular weight is considerable. Even at the temperature of liquid air, the theoretical width of each of the above three components is ·051 Å. Thus the lines 1, 2 merge into one another, so that two and not three lines are actually observed for hydrogen. The extreme

[1] H. Nagaoka, Proc. Math. Phys. Soc. Tokyo, 8, p. 237, 1915.

distance 13 is about ·2 Å., and this is of the order found by observation, which however for the reasons just given cannot be very precise.

For *any* line of the Balmer series, say the line due to the switch $n = n \rightarrow n' = 2$, possible values of k are 1, 2, 3, ... n and of k' are 1, 2. But $k' - k = \pm 1$. Therefore the possible values of k, k' are again only the pairs,

$$\left.\begin{matrix} k = 1 \\ k' = 2 \end{matrix}\right\} \quad \left.\begin{matrix} k = 2 \\ k' = 1 \end{matrix}\right\} \quad \left.\begin{matrix} k = 3 \\ k' = 2 \end{matrix}\right\};$$

higher values of k could not have a corresponding k', as k' cannot go beyond 2. This leads to three values of $\lambda - \lambda_0$, so that each Balmer line splits into three.

In the case of the Lyman series, the lines are due to the switch $n = n$ to $n' = 1$.

Possible values of k are 1, 2, 3, ... n and of k', 1.

Since $k' - k = \pm 1$, the only actual pair is

$$\left.\begin{matrix} k = 2 \\ k' = 1 \end{matrix}\right\},$$

and this leads to one line. Thus the line corresponding to the switch $n \rightarrow 1$ has

$$(\lambda - \lambda_0)/R\mu^2\lambda_0 = \left(1 - \frac{3}{4}\right) - \frac{1}{n^4}\left(\frac{n}{2} - \frac{3}{4}\right);$$

it is displaced, but not resolved.

63. *The fine structure of the lines of ionised helium.* The real experimental verification of Sommerfeld's theory lies, however, in Paschen's[1] examination of the lines of ionised helium, He_+.

This gives greater possibility of experimental measurement as R is now replaced by $4R$, and μ^2 is now $\left(\frac{2\pi Z e^2}{ch}\right)^2$, where $Z = 2$, and is thus four times as large as for hydrogen.

The line $\lambda = 4686$ Å., corresponding to a switch $n = 4$ to $n' = 3$ is resolved according to the formula

$$\frac{\nu - \nu_0}{16Rc\mu^2} = \frac{1}{81}\left(\frac{3}{k'} - \frac{3}{4}\right) - \frac{1}{256}\left(\frac{4}{k} - \frac{3}{4}\right),$$

where the μ^2 is the value for hydrogen.

[1] F. Paschen, Ann. d. Phys. **50**, p. 901, 1916.

k may be 1, 2, 3, 4 and k', 1, 2, 3, but limited by $k' - k = \pm 1$. Therefore possible pairs are

$$\left.\begin{matrix} k = 1 \\ k' = 2 \end{matrix}\right\} \quad \left.\begin{matrix} k = 2 \\ k' = 1 \end{matrix}\right\} \quad \left.\begin{matrix} k = 2 \\ k' = 3 \end{matrix}\right\} \quad \left.\begin{matrix} k = 3 \\ k' = 2 \end{matrix}\right\} \quad \left.\begin{matrix} k = 4 \\ k' = 3 \end{matrix}\right\}.$$

Thus each line is resolved into five (for, as in the Balmer series, the argument applies to all the lines).

$$\therefore \quad \frac{\lambda - \lambda_0}{16R\mu^2\lambda_0^2} = -\frac{1}{81}\left(\frac{3}{k'} - \frac{3}{4}\right) + \frac{1}{256}\left(\frac{4}{k} - \frac{3}{4}\right).$$

Also $\qquad R$ is $R_{\mathrm{He}+} = 109722$,

therefore λ_0 is given by

$$\frac{1}{\lambda_0} = 4R\left(\frac{1}{3^2} - \frac{1}{4^2}\right) = \frac{7R}{36},$$

or

$$\lambda_0 = \frac{36}{7 \times 109722} \text{ cm.} = 4686 \text{ Å}.$$

Also $\quad 16R\mu^2\lambda_0^2 = 16\left(\frac{36}{7}\right)(5{\cdot}31 \times 10^{-5})\,\lambda_0$

$$= 16\left(\frac{36}{7}\right)(5{\cdot}31 \times 10^{-5})\,(4686) \text{ Å}.$$

$$= 20{\cdot}50 \text{ Å}.$$

Therefore $\frac{\lambda - \lambda_0}{20{\cdot}5}$ has thus the following values corresponding to the following orbital switches:

$$-\frac{1}{81}\left(\frac{3}{2} - \frac{3}{4}\right) + \frac{1}{256}\left(4 - \frac{3}{4}\right) = {\cdot}0034$$

$$4_1 \to 3_2$$

$$-\frac{1}{81}\left(3 - \frac{3}{4}\right) + \frac{1}{256}\left(2 - \frac{3}{4}\right) = -{\cdot}0229$$

$$4_2 \to 3_1$$

$$-\frac{1}{81}\left(1 - \frac{3}{4}\right) + \frac{1}{256}\left(2 - \frac{3}{4}\right) = {\cdot}0018$$

$$4_2 \to 3_3$$

$$-\frac{1}{81}\left(\frac{3}{2} - \frac{3}{4}\right) + \frac{1}{256}\left(\frac{4}{3} - \frac{3}{4}\right) = -{\cdot}0070$$

$$4_3 \to 3_2$$

$$-\frac{1}{81}\left(1 - \frac{3}{4}\right) + \frac{1}{256}\left(1 - \frac{3}{4}\right) = -{\cdot}0021$$

$$4_4 \to 3_3$$

The corresponding $\lambda - \lambda_0$'s are

$$\cdot070, \quad -\cdot470, \quad \cdot037, \quad -\cdot143, \quad -\cdot043\,\text{Å}.$$

Arranged in order of λ these are

$$-\cdot470, \quad -\cdot143, \quad -\cdot043, \quad +\cdot037, \quad +\cdot070\,\text{Å}.$$

The four successive intervals between these are

$$\cdot327, \quad \cdot100, \quad \cdot080, \quad \cdot033\,\text{Å}.$$

Paschen measured the wave lengths of the five lines and found them to be

$$4685\cdot38, \quad 4685\cdot71, \quad 4685\cdot81, \quad 4685\cdot89, \quad 4685\cdot92,$$

so that his intervals are

$$\cdot33, \quad \cdot10, \quad \cdot08, \quad \cdot03,$$

in striking agreement with the theoretical values.

[For the first component of the five, we had $\lambda - \lambda_0 = -\cdot470$, and Paschen found $\lambda = 4685\cdot38$, so that $\lambda_0 = 4685\cdot85$.]

64. *The form of the relativity orbit.*

$$S = \int \frac{\partial S}{\partial r}\,dr + \int \frac{\partial S}{\partial \theta}\,d\theta$$

$$= \alpha_2\theta + \int dr\sqrt{-A + \frac{2B}{r} - \frac{C}{r^2}},$$

where

$$A = -\left(2m_0 + \frac{\alpha_1}{c^2}\right)\alpha_1, \quad B = e^2\left(m_0 + \frac{\alpha_1}{c^2}\right), \quad C = \alpha_2{}^2 - \frac{e^4}{c^2},$$

(p. 81).

By Jacobi's theory, § 43, the orbit is given by $\dfrac{\partial S}{\partial \alpha_2} = \beta_2$ and the time by $\dfrac{\partial S}{\partial \alpha_1} = t + \beta_1$.

The orbit is therefore

$$\beta_2 = \theta + \int \frac{dr}{2\sqrt{-A + \frac{2B}{r} - \frac{C}{r^2}}}\left(-\frac{2\alpha_2}{r^2}\right),$$

or

$$\theta - \beta_2 = \alpha_2 \int \frac{dr}{r^2\sqrt{-A + \frac{2B}{r} - \frac{C}{r^2}}}.$$

Writing $r = \frac{1}{u}$, this becomes

$$\theta - \beta_2 = -\alpha_2 \int \frac{du}{\sqrt{-A + 2Bu - Cu^2}},$$

or

$$\theta - \beta_2 = -\alpha_2 \int \frac{du}{\sqrt{C}\,(u_1 - u)\,(u - u_2)},$$

where

$$u_1 u_2 = \frac{A}{C}, \quad u_1 + u_2 = \frac{2B}{C}.$$

If

$$u = u_1 \sin^2 \phi + u_2 \cos^2 \phi,$$

then

$$\theta - \beta_2 = -\alpha_2 \int \frac{2d\phi}{\sqrt{C}} = \frac{-2\alpha_2 \phi}{\sqrt{C}},$$

$$\therefore \frac{\theta - \beta_2}{2} = \frac{-\alpha_2 \phi}{\sqrt{\alpha_2{}^2 - \frac{e^4}{c^2}}} = \frac{-\phi}{\sqrt{1 - \frac{e^4}{c^2 \alpha_2{}^2}}} = \frac{-\phi}{\sqrt{1 - \frac{e^4 4\pi^2}{c^2 h^2 k^2}}},$$

since

$$2\pi\alpha_2 = I_2 = n_2 h = kh. \qquad (\S\ 61)$$

$$\therefore \frac{\theta - \beta_2}{2} = \frac{-\phi}{\sqrt{1 - \frac{\mu^2}{k^2}}},$$

where $\mu = \frac{2\pi e^2}{ch}$, as before.

Writing $\gamma = \sqrt{1 - \frac{\mu^2}{k^2}}$, so that $\gamma = 1$ when relativity is neglected,

$$\left(\frac{\theta - \beta_2}{2}\right)\gamma = -\phi.$$

Hence $\frac{1}{r} = u = u_1 \sin^2 \phi + u_2 \cos^2 \phi$

$$= u_1 \sin^2 \left\{\frac{\gamma\,(\theta - \beta_2)}{2}\right\} + u_2 \cos^2 \left\{\frac{\gamma\,(\theta - \beta_2)}{2}\right\}$$

$$= \frac{u_1 + u_2}{2} + \frac{u_2 - u_1}{2} \cos \{\gamma\,(\theta - \beta_2)\},$$

and putting in the values of u_1, u_2, we obtain

$$\frac{1}{r} = \frac{B}{C} + \frac{\sqrt{B^2 - AC}}{C} \cos \{\gamma\,(\theta - \beta_2)\}$$

as the equation of the orbit.

This is of the form $\frac{l}{r} = 1 + \epsilon \cos(\gamma\theta)$,

where $$l = \frac{C}{B}, \quad \epsilon = \sqrt{1 - \frac{AC}{B^2}}, \quad \gamma = \sqrt{1 - \frac{\mu^2}{k^2}},$$

the line from which θ is measured being suitably chosen.

Thus r is a maximum if $\gamma\theta = \pi, 3\pi, \ldots$ and a minimum if $\gamma\theta = 0, 2\pi, 4\pi, \ldots$; the orbit lies between two circles of radii $l/(1 \pm \epsilon)$ and the angle between successive apse lines is

$$\pi/\gamma = \pi\bigg/\left(1 - \frac{\mu^2}{k^2}\right)^{\frac{1}{2}} = \pi\left(1 + \frac{\mu^2}{2k^2}\right) \text{ very approx.}$$

Thus the small angle between successive lines to perihelion is $2\left(\frac{\pi\mu^2}{2k^2}\right)$ or $\pi\mu^2/k^2$. If ω is the frequency in the orbit, i.e. the

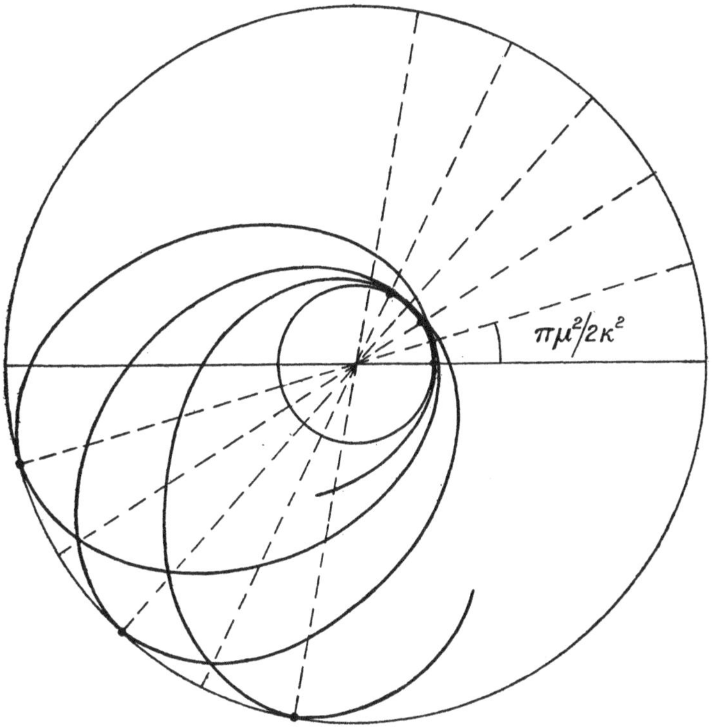

reciprocal of the time from perihelion to perihelion, then the frequency of revolution of the line to perihelion is

$$\omega \frac{\pi\mu^2}{k^2} \Big/ 2\pi \text{ or } \frac{\omega\mu^2}{2k^2}.$$

Thus there are two frequencies, one that in the orbit ω, and one that of the apse line $\omega\mu^2/2k^2$.

65. *Representation by a revolving orbit.* Consider a particle describing an orbit which revolves about the origin O in such a manner that when $POA = \theta'$, $AOX = \sigma\theta'$, where σ is a constant.

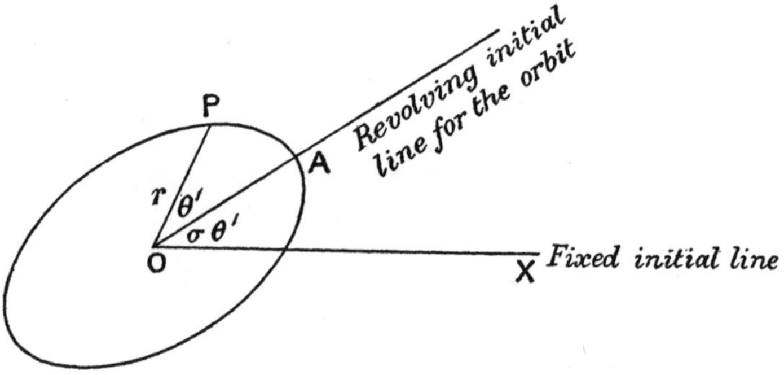

Let $\frac{1}{r} = f(\theta')$ give the form of the revolving orbit.

If θ is the *space* vectorial angle XOP, $\theta = \theta'(1 + \sigma)$, so that the path of P in space is

$$\frac{1}{r} = f\left(\frac{\theta}{1+\sigma}\right) = f(\gamma\theta),$$

where $\gamma = 1/(1 + \sigma)$.

Also when P completes a period in its orbit, so that θ' has increased by 2π, OA has completed an angle $2\pi\sigma$ or $2\pi\left(\frac{1-\gamma}{\gamma}\right)$, so that if ω is the frequency of P in its orbit, the frequency of revolution of OA is $\left(\frac{1-\gamma}{\gamma}\right)\omega$.

Thus the relativity orbit just found is the same as if the electron described a Keplerian ellipse of constant form

$$\frac{l}{r} = 1 + \epsilon \cos \theta',$$

which revolves in the manner described above (not uniformly) with frequency of revolution equal to $(1-\gamma)/\gamma$ of the frequency in the ellipse.

The frequency of revolution of the orbit is

$$\omega \left\{ \frac{1}{\gamma} - 1 \right\} = \omega \left\{ \left(1 - \frac{\mu^2}{k^2}\right)^{-\frac{1}{2}} - 1 \right\},$$

or $\omega \mu^2/2k^2$, as in § 64.

The form of the revolving ellipse is given by the equations

$$l = \frac{C}{B}, \quad \epsilon = \sqrt{1 - \frac{AC}{B^2}},$$

where approximately

$$A = -2m_0 \alpha_1, \quad B = e^2 m_0, \quad C = \alpha_2{}^2,$$

$$\therefore \; l = \frac{\alpha_2{}^2}{e^2 m_0} = \frac{k^2 h^2}{4\pi^2 m_0 e^2}.$$

If $2a$ is the major axis,

$$a = \frac{l}{1 - \epsilon^2} = \frac{C/B}{AC/B^2} = B/A = \frac{-e^2}{2\alpha_1},$$

$$\therefore \; a = \frac{e^2}{2W} = \frac{n^2 h^2}{4\pi^2 m_0 e^2}; \text{ and } b^2 = al,$$

$$\therefore \; b = \frac{nkh}{4\pi^2 m_0 e^2}. \text{ From these } b/a = k/n.$$

Thus in a 3_1 orbit, $b/a = 1/3$; in a 3_3 orbit $b/a = 1$ and the orbit is a circle.

Thus a k small compared with n means an elongated ellipse as the n_k revolving orbit.

66. *Relativity effect regarded as a central perturbation.* Consider the Newtonian law for the relativity orbit

$$\frac{l}{r} = 1 + \epsilon \cos (\gamma\theta)$$

to be described. If F is the central force required,

$$F = h'^2 u^2 \left(u + \frac{d^2 u}{d\theta^2}\right),$$

where h' is the areal constant of dynamics, and $u = 1/r$.

$$\therefore \; F = \frac{h'^2 u^2}{l} [1 + \epsilon \cos(\gamma\theta) - \epsilon\gamma^2 \cos(\gamma\theta)]$$

$$= \frac{h'^2 u^2}{l} \{1 + (1-\gamma^2)(lu - 1)\}$$

$$= \frac{h'^2 u^2}{l} \{\gamma^2 + (1-\gamma^2)\, lu\} = \frac{h'^2 \gamma^2}{l} u^2 + h'^2 (1-\gamma^2)\, u^3,$$

$$\therefore \; F = \frac{\lambda}{r^2} + \frac{\lambda'}{r^3}.$$

Thus the effect of relativity is the same as if, in the Newtonian scheme; the Coulomb field λ/r^2 were perturbed by the addition of a small field λ'/r^3. This is a particular case of Newton's theorem of the revolving orbit[1]. Also

$$\lambda'/\lambda = (1-\gamma^2)\, l/\gamma^2,$$

and using the values of § 64 this is

$$\frac{\mu^2/k^2}{1-\mu^2/k^2} l = \mu^2 l/k^2 \text{ approx.}$$

$$\therefore \; \lambda' = \frac{\mu^2}{k^2} l\lambda = \frac{\mu^2}{k^2} \left(\frac{k^2 h^2}{4\pi^2 m_0 e^2}\right) \left(\frac{e^2}{m_0}\right)$$

$$= \frac{\mu^2 h^2}{4\pi^2 m_0^{\,2}} = \frac{h^2}{4\pi^2 m_0^{\,2}} \left(\frac{2\pi e^2}{hc}\right)^2 = \frac{e^4}{m_0^{\,2} c^2}.$$

Therefore the perturbing field is $\frac{e^4}{m_0^{\,2} c^2} \cdot \frac{1}{r^3}$ per unit mass, or $\frac{e^4}{m_0 c^2} \cdot \frac{1}{r^3}$ on the electron. By regarding the relativity effect as equivalent to a small central perturbing field, Bohr[2] obtained the energy of the stationary states which determines the fine structure.

[1] I. Newton, 'Principia,' Lib. 1, Prop. 43.
[2] N. Bohr, 'Q.L.S.' Part II, p. 65.

67. *The path of an electron in a central field.* When the spectrum of a more complex atom, such as that of sodium (atomic number 11), is excited by the temperature of a flame or arc, one electron is driven to outer orbits and by its transition between stationary states produces the spectrum. These orbits of the outer electron are described under the action of the nucleus with its charge $11e$, screened by the remaining electrons (of total charge $-10e$), which still revolve comparatively close to the nucleus; the field of force in which the outer electron moves is approximately central, i.e. directed to the nucleus, but is no longer Coulombian, i.e. proportional to the inverse square of the distance.

[In the theory, relativity may be omitted as it has been seen to be equivalent to a small central field, such as the present one, in which it may be supposed included.]

If the potential energy of the system due to the field is $f(r)$, then

$$H = \tfrac{1}{2}m(\dot{r}^2 + r^2\dot{\theta}^2) + f(r) = \alpha_1,$$

and

$$p_1 = m\dot{r}, \quad p_2 = mr^2\dot{\theta},$$

so that

$$H = \frac{1}{2m}\left(p_1{}^2 + \frac{1}{r^2}p_2{}^2\right) + f(r) = \alpha_1.$$

The Hamilton-Jacobi equation is

$$\left(\frac{\partial S}{\partial r}\right)^2 + \frac{1}{r^2}\left(\frac{\partial S}{\partial \theta}\right)^2 = 2m\alpha_1 - 2mf(r);$$

as before, $\dfrac{\partial S}{\partial \theta} = \alpha_2$ and

$$\frac{\partial S}{\partial r} = \sqrt{2m\alpha_1 - 2mf(r) - \frac{\alpha_2{}^2}{r^2}}.$$

$$\therefore \; S = \alpha_2\theta + \int dr\sqrt{2m\alpha_1 - 2mf(r) - \frac{\alpha_2{}^2}{r^2}}.$$

Hence

$$I_1 = \oint \frac{\partial S}{\partial r}dr = \oint dr\sqrt{2m\alpha_1 - 2mf(r) - \frac{\alpha_2{}^2}{r^2}},$$

$$I_2 = \oint \frac{\partial S}{\partial \theta}d\theta = 2\pi\alpha_2,$$

$$\therefore \; \alpha_2 = I_2/2\pi$$

and

$$\oint dr\sqrt{2m\alpha_1 - 2mf(r) - \frac{I_2{}^2}{4\pi^2 r^2}} = I_1.$$

Thus α_1 is defined as a function of I_1 and I_2, and in general $\frac{\partial H}{\partial I_1}, \frac{\partial H}{\partial I_2}$ are different ($H \equiv \alpha_1$), so that there are two independent frequencies.

The quantum conditions are

$$I_1 = n_1 h, \quad I_2 = n_2 h;$$

the correspondence principle, exactly as in § 61, shows that n_2 can only change by unity in a transition.

The field of force may still be regarded as a perturbed Coulombian field, but with variations on a larger scale than those due to relativity. The general nature of the path will be of the 'rosette' type (figure of p. 91).

If $n_1 + n_2 = n$ and $n_2 = k$, there will be a series of stationary states such that the energy of the nth state of the kth series will be $\phi(n, k)$; the corresponding orbit will, as before, be called an n_k orbit.

Hence writing in I_1 and I_2, and putting $\alpha_1 = -W$, where W is the negative energy,

$$\oint dr \sqrt{-2mW - 2mf(r) - \frac{k^2 h^2}{4\pi^2 r^2}} = (n - k)\,h,$$

an equation which determines the energy of an n_k orbit.

CHAPTER IX

THE STARK EFFECT

68. *The Stark effect*[1]. When a stream of positive rays of *hydrogen* is subjected to a strong electric field (of order 100,000 volts-cm.) the lines of the spectrum are split up into components arranged about the original position of the line. The number of components increases with increasing frequency of the line.

If the luminous centre is observed in a direction perpendicular to the field, the components are linearly polarised, some of them in a direction at right angles to the field (r-components) and others in a direction parallel to the field (p-components).

If the luminous centre is viewed along the field, the components are unpolarised and coincide with the r-components of the previous case; no components are seen where the p-components were before.

The components are symmetrically arranged about the original line and the distances of the components from it are certain integral multiples of a small magnitude which is the same for all the lines of a given series, here the Balmer series.

69. *Theory of the effect.* The classical theory fails utterly to account for the Stark effect. The theoretical problem was first solved on Bohr's theory by Epstein[2] and by Schwartzschild[3] independently.

The dynamical problem to be solved is the motion of an electron due to a Coulomb centre of force and a constant force

[1] J. Stark, Sitzungsber. Akad. Wiss. Berlin, 1913; 'Elektrische Spektralanalyse chemischer Atome, Leipzig, 1914; Ann. d. Phys. **48**, p. 193, 1915. Also a paper on the helium lines, Ann. d. Phys. **56**, p. 569, 1918.

[2] P. Epstein, Phys. Zeitschr. **17**, p. 148, 1916; Ann. d. Phys. **50**, p. 489; **51**, p. 168, 1916.

[3] K. Schwartzschild, Berich. Akad. Berlin, p. 548, 1916.

parallel to a fixed direction. This is a particular case of two centres of force solved by Jacobi[1] by the use of elliptic coordinates. In this case one of the centres is at infinity, so giving the parallel field, and the elliptic coordinates become parabolic. The relativity separation is so small compared with the effect due to the field that it may be neglected.

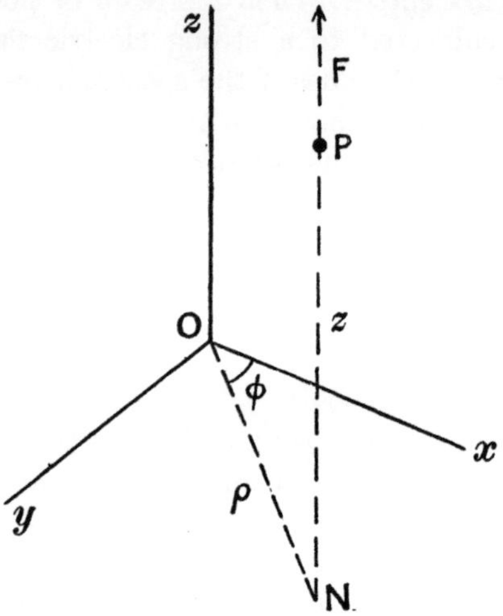

O is the nucleus and F is the electric force. The parabolic coordinates are ξ, η given by

$$\left.\begin{aligned} z &= \xi - \eta \\ \rho &= 2\sqrt{\xi\eta} \end{aligned}\right\}.$$

Eliminating η, $$z = \xi - \rho^2/4\xi,$$

or $$\rho^2 = -\,4\xi\,(z - \xi).$$

So $$\rho^2 = 4\eta\,(z + \eta).$$

Hence the curves $\xi = \text{constant}$, $\eta = \text{constant}$ are confocal parabolas, in the z, ρ plane, as shown.

The electron being $(-\,e)$, the potential energy is

$$\int\left\{\frac{e^2}{r^2}\,dr - F\,(-\,e)\,dz\right\} = -\frac{e^2}{r} + Fez.$$

[1] C. G. J. Jacobi, 'Vorlesungen über Dynamik,' p. 202.

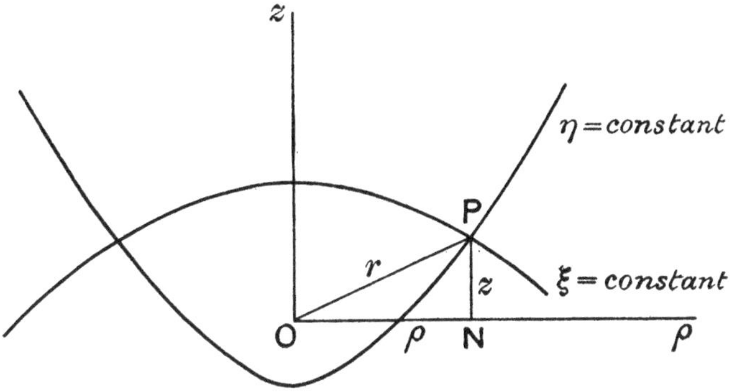

Therefore the Hamiltonian function

$$H = \tfrac{1}{2}m(\dot{z}^2 + \dot{\rho}^2 + \rho^2\dot{\phi}^2) - \frac{e^2}{r} + Fez,$$

$$\therefore\ H = \tfrac{1}{2}m\left[(\dot{\xi} - \dot{\eta})^2 + \left(\sqrt{\frac{\eta}{\xi}}.\dot{\xi} + \sqrt{\frac{\xi}{\eta}}.\dot{\eta}\right)^2 + 4\xi\eta\dot{\phi}^2\right] - \frac{e^2}{\xi + \eta} + Fe(\xi - \eta)$$

$$= \tfrac{1}{2}m\left[(\dot{\xi} - \dot{\eta})^2 + \frac{(\eta\dot{\xi} + \xi\dot{\eta})^2}{\xi\eta} + 4\xi\eta\dot{\phi}^2\right] - \frac{e^2}{\xi + \eta} + Fe(\xi - \eta),$$

$$\therefore\ H = \tfrac{1}{2}m\left\{\dot{\xi}^2\left(1 + \frac{\eta}{\xi}\right) + \dot{\eta}^2\left(1 + \frac{\xi}{\eta}\right) + 4\xi\eta\dot{\phi}^2\right\} - \frac{e^2}{\xi + \eta} + Fe(\xi - \eta).$$

Now
$$p_1 = \frac{\partial T}{\partial \dot{\xi}} = m\dot{\xi}\left(1 + \frac{\eta}{\xi}\right),$$
$$p_2 = \frac{\partial T}{\partial \dot{\eta}} = m\dot{\eta}\left(1 + \frac{\xi}{\eta}\right),$$
$$p_3 = \frac{\partial T}{\partial \dot{\phi}} = 4m\xi\eta\dot{\phi}.$$

$$\therefore\ H = \frac{1}{2m}\left[\frac{p_1^2}{1 + \frac{\eta}{\xi}} + \frac{p_2^2}{1 + \frac{\xi}{\eta}} + \frac{p_3^2}{4\xi\eta}\right] - \frac{e^2}{\xi + \eta} + Fe(\xi - \eta) = \alpha_1.$$

Therefore the Hamilton-Jacobi equation is

$$\frac{1}{2m}\left[\xi\left(\frac{\partial S}{\partial \xi}\right)^2+\eta\left(\frac{\partial S}{\partial \eta}\right)^2+\frac{1}{4}\left(\frac{1}{\xi}+\frac{1}{\eta}\right)\left(\frac{\partial S}{\partial \phi}\right)\right]^2-e^2+Fe\,(\xi^2-\eta^2)$$
$$=\alpha_1\,(\xi+\eta).$$

The variables separate.

Since H does not contain ϕ, and $\frac{\partial H}{\partial \phi}=-\dot{p}_3$, p_3 is constant.

$$\therefore\quad \frac{\partial S}{\partial \phi}=\text{constant}=\alpha_3.$$

$$\therefore\quad \xi\left(\frac{\partial S}{\partial \xi}\right)^2+\eta\left(\frac{\partial S}{\partial \eta}\right)^2+\frac{\alpha_3{}^2}{4}\left(\frac{1}{\xi}+\frac{1}{\eta}\right)-2me^2+2mFe\,(\xi^2-\eta^2)$$
$$=2m\alpha_1\,(\xi+\eta),$$

or

$$F_1\,(\xi)+F_2\,(\eta)=0.$$

Therefore each of F_1 and F_2 is a constant.

Take

$$\xi\left(\frac{\partial S}{\partial \xi}\right)^2+2mFe\xi^2-2m\alpha_1\xi-me^2+\frac{\alpha_3{}^2}{4\xi}=\alpha_2,$$

so that

$$\eta\left(\frac{\partial S}{\partial \eta}\right)^2-2mFe\eta^2-2m\alpha_1\eta-me^2+\frac{\alpha_3{}^2}{4\eta}=-\alpha_2.$$

Then

$$I_1=\oint p_1 d\xi=\oint\frac{\partial S}{\partial \xi}\,d\xi=\oint d\xi\sqrt{2m\alpha_1+\frac{me^2+\alpha_2}{\xi}-\frac{\alpha_3{}^2}{4\xi^2}-2mFe\xi},$$

$$I_2=\oint d\eta\sqrt{2m\alpha_1+\frac{me^2-\alpha_2}{\eta}-\frac{\alpha_3{}^2}{4\eta^2}+2mFe\eta},$$

$$I_3=\oint p_3 d\phi=\int_0^{2\pi}\alpha_3 d\phi=2\pi\alpha_3.$$

These integrals are of the form

$$\oint dr\sqrt{-A+\frac{2B}{r}-\frac{C}{r^2}+Dr}.$$

This is an elliptic integral; but if D is small and D^2 is negligible we can expand the square root in powers of D, so that the integral is equal to

$$\oint dr\sqrt{-A+\frac{2B}{r}-\frac{C}{r}}+\frac{1}{2}D\oint\frac{r\,dr}{\sqrt{-A+\frac{2B}{r}-\frac{C}{r^2}}}.$$

Now $$\oint \frac{r\,dr}{\sqrt{-A+\frac{2B}{r}-\frac{C}{r^2}}} = 2\int_\beta^\alpha \frac{r^2 dr}{\sqrt{A}\sqrt{(\alpha-r)(r-\beta)}},$$

where $$\alpha\beta = \frac{C}{A},$$

and $$\alpha+\beta = \frac{2B}{A}, \text{ as in § 46.}$$

Writing $$r = \alpha\sin^2\phi + \beta\cos^2\phi,$$
this becomes

$$\frac{4}{\sqrt{A}}\int_0^{\frac{\pi}{2}} (\alpha\sin^2\phi + \beta\cos^2\phi)^2 d\phi$$
$$= \frac{4}{\sqrt{A}}\left[(\alpha^2+\beta^2)\frac{\pi}{2}\cdot\frac{1.3}{2.4} + 2\alpha\beta.\frac{\pi}{16}\right]$$
$$= \frac{\pi}{4\sqrt{A}}[3(\alpha^2+\beta^2)+2\alpha\beta]$$
$$= \frac{\pi}{4\sqrt{A}}[3(\alpha+\beta)^2-4\alpha\beta]$$
$$= \frac{\pi}{A^{\frac{3}{2}}}\left(\frac{3B^2}{A}-C\right),$$

inserting the values of $\alpha+\beta$ and $\alpha\beta$.

And we have seen (§ 46) that

$$\oint dr\sqrt{-A+\frac{2B}{r}-\frac{C}{r^2}} = 2\pi\left(\frac{B}{\sqrt{A}}-\sqrt{C}\right),$$

therefore

$$\oint dr\sqrt{-A+\frac{2B}{r}-\frac{C}{r^2}+Dr}$$
$$= 2\pi\left(\frac{B}{\sqrt{A}}-\sqrt{C}\right)+\frac{\pi D}{2A}\left(\frac{3B^2}{A}-C\right), \text{ neglecting } D^2, \ldots.$$

Hence

$$I_1 = 2\pi\left(\frac{me^2+\alpha_2}{2\sqrt{-2m\alpha_1}}-\frac{\alpha_3}{2}\right) - \frac{2m\pi Fe}{2(-2m\alpha_1)^{\frac{3}{2}}}\left[\frac{3(me^2+\alpha_2)^2}{4(-2m\alpha_1)}-\frac{\alpha_3^2}{4}\right],$$

or

$$\frac{I_1}{\pi} = \frac{me^2+\alpha_2}{\sqrt{-2m\alpha_1}} - \alpha_3 + \frac{mFe}{4(-2m\alpha_1)^{\frac{3}{2}}}\left\{\frac{3(me^2+\alpha_2)^2}{2m\alpha_1}+\alpha_3^2\right\}.$$

So

$$\frac{I_2}{\pi} = \frac{me^2 - \alpha_2}{\sqrt{-2m\alpha_1}} - \alpha_3 - \frac{mFe}{4(-2m\alpha_1)^{\frac{3}{2}}}\left\{\frac{3(me^2-\alpha_2)^2}{2m\alpha_1} + \alpha_3^2\right\}$$

and

$$\frac{I_3}{\pi} = 2\alpha_3.$$

To a first approximation, neglecting F,

$$\left.\begin{aligned} \frac{I_1}{\pi} &= \frac{me^2+\alpha_2}{\sqrt{-2m\alpha_1}} - \alpha_3 \\ \frac{I_2}{\pi} &= \frac{me^2-\alpha_2}{\sqrt{-2m\alpha_1}} - \alpha_3 \\ \frac{I_3}{\pi} &= 2\alpha_3 \end{aligned}\right\},$$

$$\therefore \quad \frac{me^2+\alpha_2}{\sqrt{-2m\alpha_1}} = \frac{I_1 + \frac{1}{2}I_3}{\pi},$$

and

$$\frac{me^2-\alpha_2}{\sqrt{-2m\alpha_1}} = \frac{I_2 + \frac{1}{2}I_3}{\pi},$$

$$\therefore \quad \frac{2me^2}{\sqrt{-2m\alpha_1}} = \frac{I_1+I_2+I_3}{\pi}.$$

These values may be used in the terms containing F to obtain a result accurate to the first power of F.

$$\therefore \quad \frac{I_1}{\pi} = \frac{me^2+\alpha_2}{\sqrt{-2m\alpha_1}} - \frac{I_3}{2\pi} + \frac{mFe}{4(-2m\alpha_1)^{\frac{3}{2}}}\left[-\frac{3}{\pi^2}(I_1+\tfrac{1}{2}I_3)^2 + \frac{I_3^2}{4\pi^2}\right],$$

$$\frac{I_2}{\pi} = \frac{me^2-\alpha_2}{\sqrt{-2m\alpha_1}} - \frac{I_3}{2\pi} - \frac{mFe}{4(-2m\alpha_1)^{\frac{3}{2}}}\left[-\frac{3}{\pi^2}(I_2+\tfrac{1}{2}I_3)^2 + \frac{I_3^2}{4\pi^2}\right].$$

Adding, we have

$$\frac{I_1+I_2+I_3}{\pi} = \frac{2me^2}{\sqrt{-2m\alpha_1}} + \frac{mFe}{4(-2m\alpha_1)^{\frac{3}{2}}}\frac{3}{\pi^2}\left[(I_2+\tfrac{1}{2}I_3)^2 - (I_1+\tfrac{1}{2}I_3)^2\right]$$

$$= \frac{2me^2}{\sqrt{-2m\alpha_1}} + \frac{3mFe}{4\pi^2(-2m\alpha_1)^{\frac{3}{2}}}(I_2-I_1)(I_1+I_2+I_3),$$

and using the first approximation for $(-2m\alpha_1)$ in the last term,

$$\frac{I_1+I_2+I_3}{\pi}=\frac{2me^2}{\sqrt{-2m\alpha_1}}+\frac{3mFe}{4\pi^2}(I_2-I_1)(I_1+I_2+I_3)\left(\frac{I_1+I_2+I_3}{2\pi me^2}\right)^3,$$

$$\therefore\ \frac{2me^2}{\sqrt{-2m\alpha_1}}=\frac{I_1+I_2+I_3}{\pi}-\frac{3F}{32\pi^5m^2e^5}(I_2-I_1)(I_1+I_2+I_3)^4,$$

$$\therefore\ \frac{-2m\alpha_1}{4m^2e^4}=\frac{\pi^2}{(I_1+I_2+I_3)^2}\times\left\{1-\frac{3F}{32\pi^4m^2e^5}(I_2-I_1)(I_1+I_2+I_3)^3\right\}^{-2}$$

$$=\frac{\pi^2}{(I_1+I_2+I_3)^2}\times\left\{1+\frac{3F}{16\pi^4m^2e^5}(I_2-I_1)(I_1+I_2+I_3)^3\right\},$$

$$\therefore\ W=-\alpha_1=\frac{2\pi^2me^4}{(I_1+I_2+I_3)^2}+\frac{3F}{8\pi^2me}(I_2-I_1)(I_1+I_2+I_3)\quad\ldots\ldots(1).$$

[Similar work with a nuclear charge Ze and an electron $-e$, yields

$$W=\frac{2\pi^2mZ^2e^4}{(I_1+I_2+I_3)^2}+\frac{3F}{8\pi^2mZe}(I_2-I_1)(I_1+I_2+I_3).\Big]$$

70. *System degenerate.* The orbital frequencies are given by

$$\omega_1=\frac{\partial H}{\partial I_1}=\frac{\partial\alpha_1}{\partial I_1}=-\frac{\partial W}{\partial I_1},\ \text{etc.}$$

$$\therefore\ \omega_1=\frac{4\pi^2me^4}{(I_1+I_2+I_3)^3}-\frac{3F(I_2-I_1)}{8\pi^2me}+\frac{3F}{8\pi^2me}(I_1+I_2+I_3),$$

$$\omega_2=\frac{4\pi^2me^4}{(I_1+I_2+I_3)^3}-\frac{3F(I_2-I_1)}{8\pi^2me}-\frac{3F}{8\pi^2me}(I_1+I_2+I_3),$$

$$\omega_3=\frac{4\pi^2me^4}{(I_1+I_2+I_3)^3}-\frac{3F(I_2-I_1)}{8\pi^2me}.$$

Thus
$$\left.\begin{array}{l}\omega_1 = \omega_3 + \omega \\ \omega_2 = \omega_3 - \omega\end{array}\right\},$$
where
$$\omega = \frac{3F}{8\pi^2 me}(I_1 + I_2 + I_3).$$

Thus $\omega_1 + \omega_2 - 2\omega_3 = 0$ and the system is 'simply' degenerate (§ 55); it has three coordinates but only two independent frequencies.

71. *New 'action' variables.* We therefore change to new action and angle variables J, v instead of I, w and let the corresponding new frequencies $\dot v$ be denoted by ω'.

Write
$$\left.\begin{array}{l}I_1 + I_2 + I_3 = J_1 \\ I_2 - I_1 = J_2 \\ I_3 = J_3\end{array}\right\}.$$

Then
$$w_1 = \frac{\partial S}{\partial I_1} = \frac{\partial S}{\partial J_1}\frac{\partial J_1}{\partial I_1} + \frac{\partial S}{\partial J_2}\frac{\partial J_2}{\partial I_1} + \frac{\partial S}{\partial J_3}\frac{\partial J_3}{\partial I_1}$$
$$= v_1(1) + v_2(-1) + v_3(0).$$

Therefore $\quad w_1 = v_1 - v_2$
So $\qquad\quad w_2 = v_1 + v_2\bigg\}$,
and $\qquad\quad w_3 = v_1 + v_3$

$\therefore v_1 = \tfrac{1}{2}(w_1 + w_2)$, $\quad v_2 = \tfrac{1}{2}(w_2 - w_1)$, $\quad v_3 = w_3 - \tfrac{1}{2}(w_1 + w_2)$.

Differentiating with respect to t, we have

$\omega_1' = \tfrac{1}{2}(\omega_1 + \omega_2)$, $\quad \omega_2' = \tfrac{1}{2}(\omega_2 - \omega_1)$, $\quad \omega_3' = \omega_3 - \tfrac{1}{2}(\omega_1 + \omega_2)$,

or $\qquad\qquad \omega_1' = \omega_3$, $\quad \omega_2' = -\omega$, $\quad \omega_3' = 0$,

since $\qquad\qquad \omega_1 + \omega_2 - 2\omega_3 = 0$.

[Note that $\dot v_3 = \omega_3' = 0$, $\;\therefore\; v_3$ is a constant.]

Thus corresponding to the new action variables J_1, J_2 are new frequencies ω_1', ω_2' or numerically ω_3, ω.

The former is approximately the frequency in the undisturbed Keplerian orbit and the latter is a new frequency of magnitude $\dfrac{3FJ_1}{8\pi^2 me}$.

The negative energy $W = \dfrac{2\pi^2 me^4}{J_1^2} + \dfrac{3FJ_1 J_2}{8\pi^2 me}$, from (1) p. 103.

We now have W expressed in terms of two action variables which correspond to two independent frequencies; the quantum conditions are therefore $J_1 = nh$, $J_2 = kh$, where n, k are integers.

Hence for an n_k orbit

$$W = \frac{2\pi^2 me^4}{n^2 h^2} + \frac{3Fh^2 nk}{8\pi^2 me}.$$

72. *Application of the correspondence principle.* The possible transitions and the polarisation of the lines are determined by the correspondence principle.

On p. 100 were found results of the form

$$\left.\begin{aligned} I_1 &= f_1(\xi, \alpha_1, \alpha_2, \alpha_3) \\ I_2 &= f_2(\eta, \alpha_1, \alpha_2, \alpha_3) \\ I_3 &= 2\pi\alpha_3 \end{aligned}\right\} \quad \ldots\ldots\ldots\ldots\ldots\ldots(1),$$

Also
$$\begin{aligned} S &= \int \frac{\partial S}{\partial \phi} d\phi + \int \frac{\partial S}{\partial \xi} d\xi + \int \frac{\partial S}{\partial \eta} d\eta \\ &= \alpha_3\phi + \int d\xi \text{ (function of } \xi, \alpha_1, \alpha_2, \alpha_3) \\ &\qquad + \int d\eta \text{ (function of } \eta, \alpha_1, \alpha_2, \alpha_3) \\ &= \alpha_3\phi + \psi_1(\xi, \alpha_1, \alpha_2, \alpha_3) + \psi_2(\eta, \alpha_1, \alpha_2, \alpha_3). \end{aligned}$$

From (1), the α's can be expressed in terms of the I's and ξ, η, so that

$$S = \frac{I_3\phi}{2\pi} + \psi(\xi, \eta, I_1, I_2, I_3),$$

$$\left.\begin{aligned} \therefore\ w_1 &= \frac{\partial S}{\partial I_1} = F_1(\xi, \eta, I_1, I_2, I_3) \\ w_2 &= \frac{\partial S}{\partial I_2} = F_2(\xi, \eta, I_1, I_2, I_3) \\ w_3 &= \frac{\partial S}{\partial I_3} = \frac{\phi}{2\pi} + F_3(\xi, \eta, I_1, I_2, I_3) \end{aligned}\right\} \quad \ldots\ldots\ldots(2).$$

From the first pair of (2), ξ, η are functions of w_1, w_2 and the I's, and therefore ρ, z are functions of w_1, w_2 and the I's.

Therefore from the general theory of angle variables, ρ, z are periodic in w_1, w_2 of period 1.

From the third of (2)

$$\phi = 2\pi w_3 - 2\pi F_3(\xi, \eta, I_1, I_2, I_3)$$
$$= 2\pi w_3 + F(w_1, w_2, I_1, I_2, I_3).$$

Therefore $\rho e^{i\phi}$ is of the form

$$e^{2\pi i w_3}\Phi(w_1, w_2, I_1, I_2, I_3).$$

Hence z is of the form

$$\Sigma A e^{2\pi i(\tau_1 w_1 + \tau_2 w_2)},$$

and $\rho e^{i\phi}$ ($\equiv x + iy$) is of the form

$$\Sigma B e^{2\pi i(\tau_1 w_1 + \tau_2 w_2 + w_3)}.$$

In terms of the new angle variables v_1, v_2, v_3 (of which v_3 is a constant) we have, using

$$w_1 = v_1 - v_2, \quad w_2 = v_1 + v_2, \quad w_3 = v_1 + v_3,$$

that z is of the form $\Sigma A e^{2\pi i[(\tau_1+\tau_2)v_1 + (\tau_2-\tau_1)v_2]}$,
and $x + iy$ of the form

$$\Sigma B e^{2\pi i[(\tau_1+\tau_2+1)v_1 + (\tau_2-\tau_1)v_2 + v_3}$$
$$= \Sigma B' e^{2\pi i[(\tau_1+\tau_2+1)v_1 + (\tau_2-\tau_1)v_2]},$$

since v_3 is a constant; also since

$$v_1 = \omega_1' t + \text{const., and } v_2 = \omega_2' t + \text{const.},$$

we have that z is of the form

$$\Sigma A' e^{2\pi i(\tau_1'\omega_1' + \tau_2'\omega_2')t},$$

where $\tau_1' + \tau_2' = (\tau_1 + \tau_2) + (\tau_2 - \tau_1) = 2\tau_2$, and is *even*,
and $x + iy$ is of the form

$$\Sigma B'' e^{2\pi i(\tau_1'\omega_1' + \tau_2'\omega_2')t},$$

where $\tau_1' + \tau_2' = (\tau_1 + \tau_2 + 1) + (\tau_2 - \tau_1) = 2\tau_2 + 1$,
and is *odd*.

Now the J_1, J_2 correspond to v_1, v_2 and so to ω_1', ω_2' and $J_1 = nh$, $J_2 = kh$; so that the τ_1', τ_2' of the Fourier series correspond to transitions Δn, Δk of Bohr's theory, by the principle of correspondence.

Thus the z component of the motion is associated with transitions such that $\Delta n + \Delta k$ is *even*, and the motion in x and y with transitions such that $\Delta n + \Delta k$ is *odd*.

On the classical theory, the z motion represents a polarisation parallel to the field and the x, y motion a circular polarisation in a plane at right angles to the field; by the principle of correspondence we infer that transitions for which $\Delta n + \Delta k$ is *even* lead to lines linearly polarised parallel to the field, and those for which $\Delta n + \Delta k$ is *odd* lead to lines circularly polarised in a plane at right angles to the field.

73. Again I_1, I_2, I_3 may take any values between 0 and J_1, since $J_1 = I_1 + I_2 + I_3$; hence $J_2 \{\equiv (I_2 - I_1)\}$ may take any values between $-J_1$ and $+J_1$. But if, for example, it takes the value J_1, then $I_2 = J_1$ and $I_1 = 0$; and if it takes the value $-J_1$, then $I_2 = 0$ and $I_1 = J_1$. In each of these cases $I_3 = 0$.

Now $I_3 = 0$ means that $\alpha_3 = 0$ or $p_3 = 0$, i.e. the angular momentum about the axis of z is zero. Therefore the orbit is in a plane which passes through Oz. Now the libration limits of ξ, η are given by the roots of equations of the form

$$-A + \frac{2B}{r} - \frac{C}{r^2} = 0 \text{ (p. 101)},$$

where for both ξ and η, $\quad C = \frac{\alpha_3^2}{4}$.

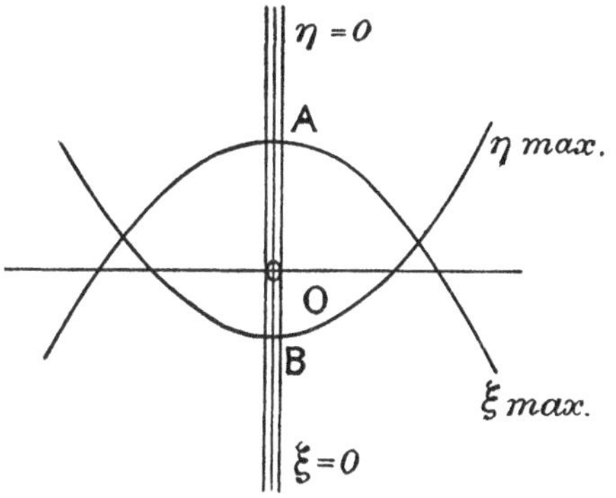

Hence in the above case $C = 0$ and one of the values of ξ and of η at the libration limits is zero; the path would in time touch the parabolas $\xi = 0$ and $\eta = 0$ at all points from A to B and so must at some time pass through the nucleus. These are the 'pendelbahnen' of Epstein, and must be excluded from the possible set of stationary states. This means that $|J_2|$ may take all values less than J_1, but it may not $= J_1$.

Thus $|k|$ must be $< n$, for $J_1 = nh$, $J_2 = kh$. Therefore k can take the values

$$n-1,\ n-2, \ldots 2, 1, 0, -1, -2 \ldots -(n-2),\ -(n-1)$$

but not the values $\pm n$. These last as we have seen would involve in time a collision with the nucleus.

74. *Numerical results of the theory.* For a transition from an n_k orbit to an $n'_{k'}$ orbit, the frequency ν of the line emitted is given by

$$h\nu = E - E' = W' - W.$$

$$\therefore\ h\nu = \frac{2\pi^2 me^4}{h^2}\left(\frac{1}{n'^2} - \frac{1}{n^2}\right) + \frac{3Fh^2}{8\pi^2 me}(n'k' - nk).$$

For the normal undisturbed line of frequency ν_0,

$$h\nu_0 = \frac{2\pi^2 me^4}{h^2}\left(\frac{1}{n'^2} - \frac{1}{n^2}\right),$$

$$\therefore\ \nu - \nu_0 = \frac{3Fh}{8\pi^2 me}(n'k' - nk) = \Delta\,(n'k' - nk),$$

where $$\Delta = \frac{3Fh}{8\pi^2 me}.$$

Consider the H_α line, where the transition is from $n = 3$ to $n' = 2$

$$\frac{\nu - \nu_0}{\Delta} = 2k' - 3k,$$

and from the theory of § 73, k' may be $1, 0, -1$ and k may be $2, 1, 0, -1, -2$.

From these we have the table:

n	3														
n'	2														
k	2			1			0			-1			-2		
k'	1	0	-1	1	0	-1	1	0	-1	1	0	-1	1	0	-1
$2\kappa' - 3\kappa$	-4	-6	-8	-1	-3	-5	2	0	-2	5	3	1	8	6	4
$k+n$	5	5	5	4	4	4	3	3	3	2	2	2	1	1	1
$k'+n'$	3	2	1	3	2	1	3	2	1	3	2	1	3	2	1
$(k'+n')-(k+n)$	-2	-3	-4	-1	-2	-3	0	-1	-2	1	0	-1	2	1	0
Polarisation	p	r	p	r	p	r	p	r	p	r	p	r	p	r	p

p, r denote linear polarisation parallel, at right angles to the field.

Since $\nu - \nu_0 = \Delta\,(2k' - 3k)$, the table shows that

$$\nu - \nu_0 = \pm\, n\Delta,$$

where n may be any number from 0 to 8, excluding 7.

Thus the H_α line yields 15 components equally spaced at intervals Δ, except the two outer ones where the space is 2Δ. The lines are polarised as indicated.

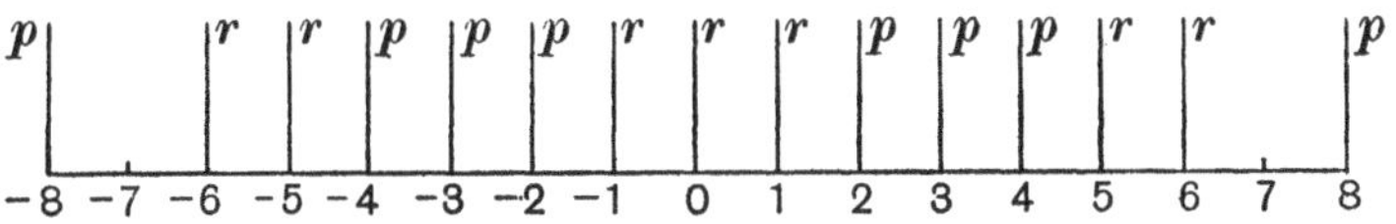

From the theory, if $\Delta n + \Delta k$ is even there is linear polarisation parallel to the field, and if $\Delta n + \Delta k$ is odd there is circular polarisation in a plane at right angles to the field. The values of $\Delta n + \Delta k$ or $(k' + n') - (k + n)$ are shown in the table and the corresponding polarisations in the row beneath.

When the lines are viewed at right angles to the field, that is to Oz, the lines p in the table are linearly polarised parallel to the field; the lines r, being circularly polarised in the plane x, y and seen edgewise, give linear polarisation at right angles to the field.

If the lines are viewed along the field, the lines p corresponding to a vibration along the line of vision are not seen, and the lines r are seen without polarisation because, owing to the symmetry of the field about Oz, the number of transitions giving light circularly polarised in a positive sense and the number giving light circularly polarised in a negative sense will in the mean be the same.

Thus in the former case, the photographs of the p lines and of the r lines, which can be separately taken by the usual experimental methods for polarised light, would be of the type shown in the figures on the next page. The lengths of the lines indicate the intensity of the lines as observed; only the middle six of the upper set and the middle three of the lower were shown in Stark's photographs; the relative spacing of the lines agreed with this theory.

The calculation of the relative intensities of the lines is long and difficult, and as the theory stands at present a mean has to be struck between the coefficients of the harmonic terms in the two orbits (concerned in a transition) which determine the intensity of a given line. Kramers[1] has however obtained results in very considerable agreement with the experimental data.

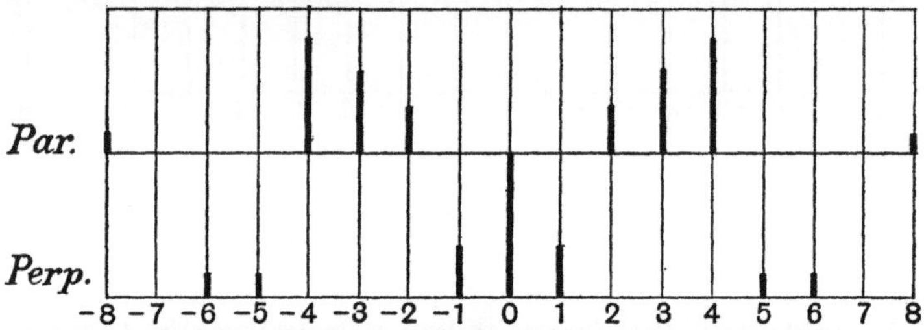

It should be noticed that Δ is independent of n or n' and is the same therefore for all lines of the Balmer series.

The number of possible components is equal to the product of the numbers of possible values of k and k', i.e.

$$(2n - 1)(2n' - 1).$$

For the Balmer line $3 \rightarrow 2$ (H_α) the number of components is $5 \times 3 = 15$;

For the Balmer line $4 \rightarrow 2$ (H_β) the number of components is $7 \times 3 = 21$;

For the Balmer line $5 \rightarrow 2$ (H_γ) the number of components is $9 \times 3 = 27$;

and so on.

75. *For helium*, $\Delta = \frac{3Fh}{8\pi^2 mZe}$, where $Z = 2$, and is one-half its value for hydrogen. For the 4686 line, the transition is from $n = 4$ to $n' = 3$, so that there are 7×5, or 35 components.

Interesting diagrams both for hydrogen and helium are given at the end of Kramers' dissertation just referred to (Plates I–IV).

[1] H. A. Kramers, 'Intensities of Spectral Lines,' D. Kgl. Danske Vidensk. Selsk. Skrifter, 8. iii. 3, p. 287, 1919.

The actual displacement Δ for a field of 100,000 volts-cm. is

$$\frac{3Fh}{8\pi^2 me} = \frac{3 \times \frac{100{,}000}{300} \times 6{\cdot}55 \times 10^{-27}}{8\pi^2 (9{\cdot}04 \times 10^{-28})(4{\cdot}77 \times 10^{-10})}$$

$$= \frac{6{\cdot}55}{8\pi^2 (9{\cdot}04)(4{\cdot}77)} 10^{14}, \text{ in frequency.}$$

If it is expressed in wave length, then since

$$\nu = \frac{c}{\lambda}, \quad |\Delta\nu| = \frac{c}{\lambda^2} |\Delta\lambda|$$

and the displacement in wave length

$$= \frac{\lambda^2}{c} |\Delta\nu|,$$

for the H_α line this

$$= \frac{6{\cdot}55 \times 10^{14}}{8\pi^2 (9{\cdot}04)(4{\cdot}77)} \times \frac{(6562{\cdot}8 \times 10^{-8})^2}{3 \times 10^{10}}$$

$$= 2{\cdot}8 \times 10^{-8} \text{ cm.}$$

$$= 2{\cdot}8 \text{ Å.}$$

Thus the extreme separation, 16Δ is 44·8 Å.; the extreme lines so far observed, as the outer ones have small intensity, correspond to $\pm 4\Delta$ or a range $8\Delta = 22{\cdot}4$ Å. It is thus apparent how large this effect is compared with the relativity separation for H_α which was only ·2 Å.

CHAPTER X

THE ZEEMAN EFFECT

76. *The Zeeman effect.* In 1897, Zeeman[1] discovered that a strong magnetic field splits up the *hydrogen* lines into components. When viewed at right angles to the field there were three components; the centre one coinciding with the unresolved line and polarised parallel to the field; the outer ones equidistant from the centre one and polarised at right angles to the field. When viewed along the field there were only two components, in the same positions as the outer ones of the previous case but circularly polarised in opposite directions.

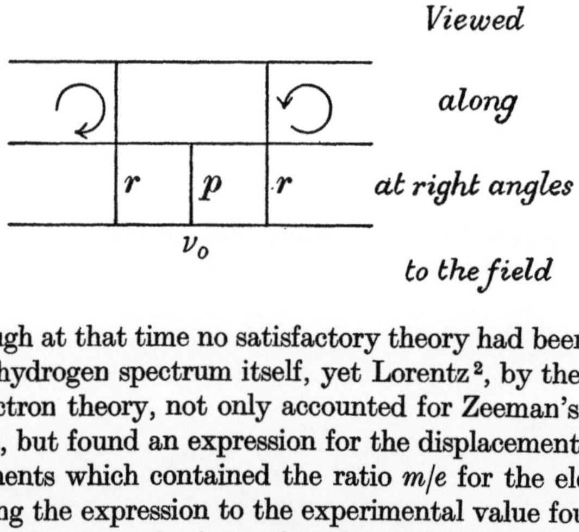

Though at that time no satisfactory theory had been given of the hydrogen spectrum itself, yet Lorentz[2], by the use of the electron theory, not only accounted for Zeeman's observations, but found an expression for the displacement of the components which contained the ratio m/e for the electron. Equating the expression to the experimental value found by Zeeman, the ratio of m/e was found to be in close agreement with the value obtained directly from cathode rays.

[1] P. Zeeman, Phil. Mag. **43**, p. 226, 1897; Collected Papers on Magneto-optical phenomena, Leiden, 1921.

[2] H. A. Lorentz, 'The Theory of Electrons,' chap. III, 1909.

This result was accepted as convincing confirmation of the new electron theory and showed too that the origin of the spectrum itself was the motion of electrons within the atom

An important contribution to the theory was the general theorem obtained by Larmor[1], which includes Lorentz's results, and has since been used by Bohr in treating the Zeeman effect on the lines of his new theory.

Larmor showed that the effect of a uniform magnetic field on the motion of an electron under any forces is the same, to a first approximation, as if the magnetic field were absent and the whole system had a uniform rotation about an axis parallel to the direction of the field with a frequency ω_H, where $\omega_H = eH/4\pi mc$, and H is the intensity of the field.

77. *Proof of Larmor's theorem.* If X, Y, Z is the field of force on the electron and H is the constant magnetic field, taken parallel to Oz, where Ox, Oy, Oz are fixed axes, the equations of motion are

$$\left.\begin{aligned} m\ddot{x} &= \frac{e}{c}H\dot{y} + X \\ m\ddot{y} &= -\frac{e}{c}H\dot{x} + Y \\ m\ddot{z} &= Z \end{aligned}\right\}, \text{ where } e \text{ is in E.S.U.}$$

$$\therefore \left.\begin{aligned} m(\ddot{x} - \lambda\dot{y}) &= X \\ m(\ddot{y} + \lambda\dot{x}) &= Y \\ m\ddot{z} &= Z \end{aligned}\right\}, \text{ where } \lambda = \frac{eH}{mc} \quad \text{.........(1).}$$

If now the electron is supposed free from the action of the magnetic field and is referred to axes rotating round Oz with constant angular velocity Ω, its velocities are

$$u = \dot{x} - y\Omega, \quad v = \dot{y} + x\Omega, \quad w = \dot{z},$$

and its accelerations

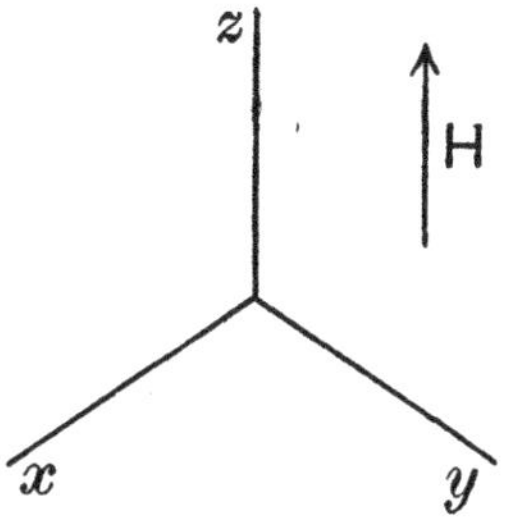

$$\dot{u} - v\Omega, \quad \dot{v} + u\Omega, \quad \dot{w},$$

or $$\ddot{x} - 2\dot{y}\Omega - x\Omega^2, \quad \ddot{y} + 2\dot{x}\Omega - y\Omega^2, \quad \ddot{z},$$

[1] J. Larmor, 'Aether and Matter,' p. 341, Cambridge, 1900.

and the equations of motion are

$$\left.\begin{aligned} m(\ddot{x} - 2\Omega\dot{y} - \Omega^2 x) &= X \\ m(\ddot{y} + 2\Omega\dot{x} - \Omega^2 y) &= Y \\ m\ddot{z} &= Z \end{aligned}\right\} \quad \text{...........(2).}$$

If $\lambda = 2\Omega$, and therefore Ω^2 is negligible, the equations (2) referred to the moving axes *without* the magnetic field are the same as those (1) for fixed axes *with* the field.

Thus the effect of the magnetic field is simply to impose a rotation with angular velocity $\Omega = \frac{\lambda}{2} = \frac{eH}{2mc}$ around its axis on some undisturbed motion of the system. The *frequency* ω_H corresponding to the angular velocity Ω is

$$\omega_H = \frac{\Omega}{2\pi} = \frac{eH}{4\pi mc}.$$

[This result requires the specification of X, Y, Z to be the same with regard to both sets of axes; this would, for example, be the case for a central field acting on the electron.]

78. *The theory of the Zeeman effect* in terms of the quantum ideas was first given independently by Sommerfeld[1] and by Debye[2].

By the dynamical theory of § 37

$$L = T - V - \frac{e}{c}(\dot{x}F + +),$$

and for a uniform magnetic field γ parallel to Oz, $F = -\frac{1}{2}\gamma y$, $G = \frac{1}{2}\gamma x$, $H = 0$, so that for a hydrogen atom,

$$L = T + \frac{e^2}{r} - \frac{\gamma e}{2c}(x\dot{y} - y\dot{x}),$$

and the Hamiltonian function

$$= T + V = T - \frac{e^2}{r}.$$

In space polar coordinates r, θ, ϕ,

$$L = \tfrac{1}{2}m(\dot{r}^2 + r^2\dot{\theta}^2 + r^2\sin^2\theta\dot{\phi}^2) + \frac{e^2}{r} - \frac{\gamma e}{2c} r^2 \sin^2\theta\dot{\phi},$$

$$H = \tfrac{1}{2}m(\dot{r}^2 + r^2\dot{\theta}^2 + r^2\sin^2\theta\dot{\phi}^2) - \frac{e^2}{r}.$$

[1] A. SOMMERFELD, Phys. Zeitschr. **17**, p. 491, 1916.
[2] P. DEBYE, Phys. Zeitschr. **17**, p. 507, 1916.

By Lagrange's equations (p. 50),

$$p_1 = \frac{\partial L}{\partial \dot{r}} = m\dot{r}, \quad p_2 = \frac{\partial L}{\partial \dot{\theta}} = mr^2\dot{\theta},$$

$$p_3 = \frac{\partial L}{\partial \dot{\phi}} = mr^2 \sin^2\theta \left(\dot{\phi} - \frac{\gamma e}{2mc}\right),$$

so that

$$H = \frac{1}{2m}\left[p_1{}^2 + \frac{p_2{}^2}{r^2} + \frac{1}{r^2\sin^2\theta}\left(p_3 + r^2\sin^2\theta\,\frac{\gamma}{2\epsilon}\right)^2\right] - \frac{e^2}{r}.$$

The method of the Hamilton-Jacobi equation, with separation of variables, was then applied to this function.

The theory however can be developed more simply, and perhaps with more physical significance, on the lines indicated by Bohr, in which he makes a direct use of Larmor's theorem.

79. *Bohr's[1] theory of the Zeeman effect.* By Larmor's theorem, the motion of the electron in the hydrogen atom is the motion in a Keplerian ellipse which it would have if there were no field, together with a uniform rotation of the ellipse with frequency ω_H about the axis of the field, where $\omega_H = eH/4\pi mc$.

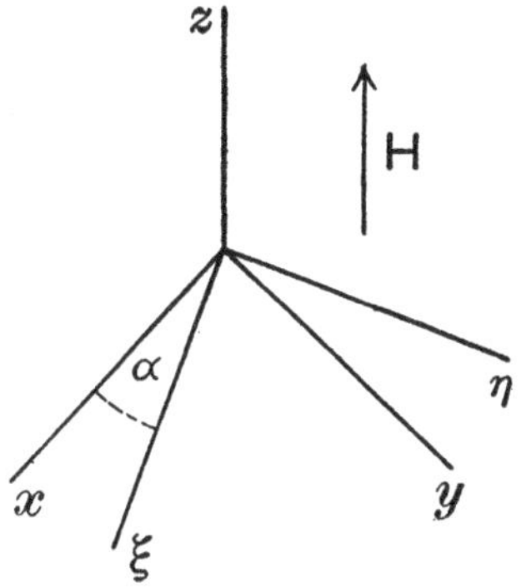

Thus if x, y, z are fixed axes and ξ, η, z revolve with angular velocity $2\pi\omega_H$, the angle α is $\pm\, 2\pi\omega_H t$ and the electron moves with frequency ω in an ellipse fixed with respect to ξ, η, z.

Hence ξ, η, z are periodic with this frequency, and are each of the form

$$\Sigma_\tau A e^{2\pi i(\tau\omega)t}.$$

But
$$x + iy = e^{i\alpha}(\xi + i\eta)$$
$$= e^{\pm 2\pi i\omega_H t}(\xi + i\eta),$$

[1] N. Bohr, 'Q.L.S.' Part II, p. 79; Part I, p. 23.

so that $x + iy$ is of the form

$$\left.\begin{array}{ll} & \Sigma B e^{2\pi i(\tau\omega \pm \omega_H)t} \\ \text{and } z \text{ is of the form} & \Sigma A e^{2\pi i(\tau\omega)t} \end{array}\right\} \quad \ldots\ldots\ldots\ldots\ldots\ldots(1).$$

The motion of the atom in the field is thus a doubly periodic motion with the fundamental frequencies ω, ω_H. If I, I_H are the action variables corresponding to these frequencies, the energy difference of two neighbouring motions is $\delta E = \omega\delta I + \omega_H \delta I_H$ (§ 53), and the stationary states will be subject to the conditions

$$I = nh, \quad I_H = n_H h.$$

Hence for the total energy of the atom as a function of I and I_H, we have, using

$$\omega = \frac{4\pi^2 e^4 m}{I^3} \text{ from p. 47,}$$

and $$\omega_H = eH/4\pi mc,$$

that $$E = -\frac{2\pi^2 e^4 m}{I^2} + \frac{eH}{4\pi mc} I_H,$$

and the condition $\omega I + \omega_H I_H = \bar{A}$ is satisfied (§ 54).

Writing $I = nh$, $I_H = n_H h$, the energy in the stationary states is

$$E = -\frac{2\pi^2 e^4 m}{h^2} \cdot \frac{1}{n^2} + \frac{heH}{4\pi mc} n_H \quad \ldots\ldots\ldots\ldots(2).$$

For the frequency of the line due to a change $n \rightarrow n'$, $n_H \rightarrow n_H'$,

$$h\nu = E - E'$$

or, $$\nu = \frac{2\pi^2 e^4 m}{h^3}\left(\frac{1}{n'^2} - \frac{1}{n^2}\right) + \frac{eH}{4\pi mc}(n_H - n_H') \ldots\ldots(3).$$

If ν_0 is the frequency corresponding to the change $n \rightarrow n'$ in the absence of the field,

$$\nu - \nu_0 = \frac{eH}{4\pi mc}(n_H - n_H').$$

The correspondence principle shows from (1) that while Δn may have any value, Δn_H can only be ± 1 or 0. The former ($\Delta n_H = \pm 1$) corresponds to circular rotations parallel to the plane xy and the latter ($\Delta n_H = 0$) to linear harmonic vibrations parallel to Oz.

In the former case

$$\nu - \nu_0 = \frac{eH}{4\pi mc}(\pm 1) = \pm \frac{eH}{4\pi mc}$$

and in the latter case

$$\nu - \nu_0 = \frac{eH}{4\pi mc}(0) = 0.$$

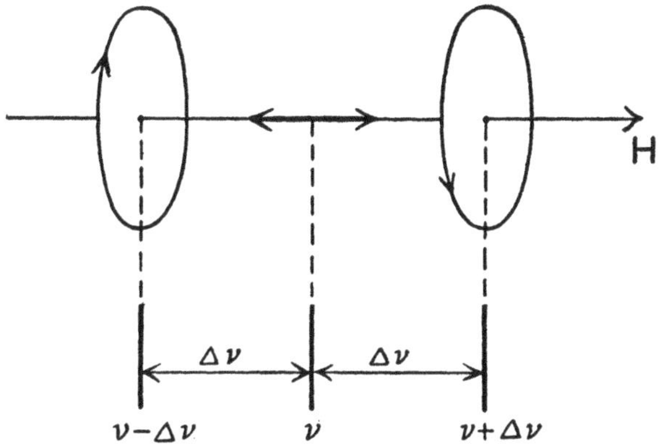

The figure shows the corresponding lines and their polarisations, and $\Delta\nu = eH/4\pi mc$.

When viewed transversely (at right angles to H) the circular rotations (seen edgewise) appear to be linear oscillations so that the two outer lines are seen polarised at right angles to H; the centre one is polarised parallel to H. When viewed longitudinally (parallel to H) the outer lines are circularly polarised in opposite directions, and the centre one is not seen as it is a vibration along the line of vision.

Thus in the former case the line is split up into a triplet, and in the latter into a doublet, with polarisations as described at the beginning of the chapter.

It should be noticed that in deducing the frequency difference $\Delta\nu$ from equation (2), Planck's constant h divided out, so that the displacement $\Delta\nu$ is independent of h, while in the theory of the Stark effect the separation of the lines was given by an expression containing h. Thus it is seen how it is that the classical theory, in which the discontinuity h played no part, could readily account for the Zeeman effect, but fail at the Stark effect.

80. *Magnitude of the effect.* $\Delta\nu = eH/4\pi mc$, and using the known values of e, m, c (§ 14), $\Delta\nu = 14 \times 10^5 H$, or in wave number

$$\Delta N = \Delta\left(\frac{\nu}{c}\right) = 4{\cdot}7 \times 10^{-5}\, H \text{ cm.}^{-1}.$$

CHAPTER XI

THE SERIES SPECTRA OF THE ELEMENTS

81. *The K, L, M, ... groups.* In elements of higher atomic number the orbits of the electrons can be classified into groups; the orbits of each group have the same energy, if to a first approximation the multiple structure of the spectral lines is neglected. For example, the familiar yellow D line of sodium is really two of wave lengths 5890 and 5896 Å., but for the moment will be considered as a single line.

The first group (the K group or 'shell'), reckoning from the nucleus outwards, contains, when complete, two electrons, each in a separate orbit whose quantum number is 1; the atom with this group just complete is the helium atom (atomic number 2). The next group (the L group) contains, when complete, eight electrons in orbits of quantum number 2; the atom with the K group and L group just complete is the neon atom (atomic number 10). The building up of the L shell during which the number of electrons in the shell increases from 1 to 8 corresponds to the formation of the atoms from lithium (atomic number 3) to neon (atomic number 10).

The formation of the M shell with its eight electrons in orbits of quantum number 3 leads from sodium (atomic number 11) to argon (atomic number 18).

A further addition of one electron commences the N shell, with potassium (atomic number 19) and one more electron leads to calcium (atomic number 20), but beyond this point the building up of the atoms ceases to continue in this regular manner by the addition of electron after electron to the outermost shell; it is found that for the next element, scandium (atomic number 21), the addition of the new electron does not lead to the numbers 2, 8, 8, 3 of electrons in the *K*, *L*, *M*, *N* shells, but that a redistribution occurs to the more stable state of 2, 8, 9, 2.

The further stages of this process of atom building, with spectral evidence to justify it, will be considered in detail later (chapter XVI); the above will suffice for the purpose of this chapter.

The element of greatest atomic number known at present, uranium (atomic number 92), contains seven groups of electrons. Outside these occupied orbits (which have quantum numbers ranging from 1 to 7 in the case of uranium) are other possible stationary states not occupied by electrons ('virtual' orbits) until the virtual orbit corresponding to $n \to \infty$ is reached, the periphery of the atom.

82. *Optical spectra.* When the flame or the electric arc is used to excite a spectrum, one electron of the outermost shell is driven out from its normal orbit to one of the 'virtual' orbits. Such an orbit is large compared with the distances of the other electrons from the nucleus, and by spontaneous transitions to orbits of lesser energy the spectrum is emitted. For example, in the case of sodium (atomic number 11) the outer electron is acted upon by the nucleus ($+ 11e$) shielded by the remaining ten electrons ($- 10e$); so that the force on the outer electron in the parts of its orbit whose distance is large compared with the dimensions of the inner 'core' (the nucleus and 10 electrons) is effectively a Coulombian one due to a charge $+ e$, as in the hydrogen atom; but in the parts of the orbit nearer the nucleus, the field is approximately central, but not Coulombian, so that the orbit is of the 'rosette' type, as shown in § 67, the greater part of the orbit being approximately a Keplerian ellipse in form. The figure on the page opposite indicates the type of orbit.

The negative energy W in one of these orbits was shown in § 67 to depend upon two quantum numbers n, k, so that $W = \phi(n, k)$. As, for large values of n where the whole orbit is very distant from the nucleus, the orbit is the same as a hydrogen orbit, $\phi(n, k)$ must tend to Rhc/n^2 as $n \to \infty$.

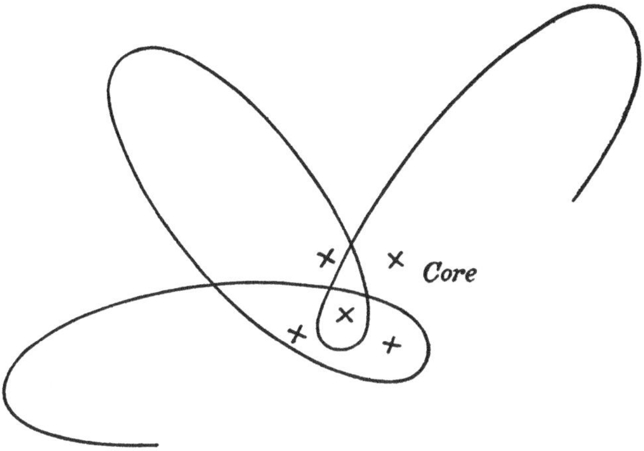

Hence $W = \frac{Rhc}{n^2} f(n, k)$, where $f(n, k) \to 1$ as $n \to \infty$.

Thus there is not, as with hydrogen, a single series of stationary states, but a set of such series, the negative energy of the atom in the nth state of the kth series being $\frac{Rhc}{n^2} f(n, k)$.

83. *The Rydberg and Ritz formulae.* Rydberg[1] proposed the equivalent of the formula $W = \frac{Rhc}{(n - \alpha_k)^2}$, where α_k is constant for a given k; while Ritz[2] gave the equivalent of

$$W = Rhc/(n - \alpha_k - \beta_k W)^2,$$

where α_k, β_k are constant for a given k.

Using the approximation $W = Rhc/n^2$ for W on the right-hand side of Ritz's formula, we obtain the formula

$$\frac{Rhc}{\left(n - \alpha_k - \frac{\gamma_k}{n^2}\right)^2},$$

used by Hicks[3].

[1] J. R. Rydberg, K. Svenska Akad. Handl. **23**, 1889.

[2] W. Ritz, Ann. d. Phys. **12**, p. 264, 1903; Phys. Zeitschr. **9**, p. 521, 1908; Gesammelte Werke, Paris, 1911.

[3] W. M. Hicks, 'The Analysis of Spectra,' Cambridge, 1922.

These are all of the above type $\frac{Rhc}{n^2} f(n, k)$, where $f(n, k) \to 1$ as $n \to \infty$, or $W = \phi(n, k)$.

Thus the possible energies for the series of orbits is in general

$$\left.\begin{array}{r} \phi(1, 1), \quad \phi(2, 1), \quad \phi(3, 1), \quad \phi(4, 1) \ldots \\ \text{for the series of states } k = 1 \ldots S \\ \phi(2, 2), \quad \phi(3, 2), \quad \phi(4, 2) \ldots \\ \text{for the series of states } k = 2 \ldots P \\ \phi(3, 3), \quad \phi(4, 3) \ldots \\ \text{for the series of states } k = 3 \ldots D \\ \phi(4, 4) \ldots \\ \text{for the series of states } k = 4 \ldots B \end{array}\right\} (\alpha).$$

These orbital states are known as the *s*, *p*, *d*, *b* 'terms.'

84. *Spectral series.* The theory of the central field (§ 67) showed, by the use of the correspondence principle, that k can only change by one unit, so that the spectral lines can only correspond to transitions between two energy levels or 'terms' in *consecutive* rows of the table (α).

Transitions from the P row to the first term of the S row give the *principal series* of spectral lines of the element.

Those from the S row (except the first term) to the first term of the P row give the *sharp series*.

Transitions from the D row to the first term of the P row give the *diffuse series*.

Those from the B row to the first term of the D row give the *Bergmann series*.

It should be noticed that in the case of an element like sodium which in the normal state has its outer electron in the N group (quantum number 3), n cannot be less than 3, so that the energy terms would be

$$\left.\begin{array}{lllll} \phi(3, 1), & \phi(4, 1) & \phi(5, 1) & \phi(6, 1) & \text{for the } S \text{ row} \\ & \phi(3, 2) & \phi(4, 2) & \phi(5, 2) & \quad ,, \quad P \; ,, \\ & & \phi(3, 3) & \phi(4, 3) & \quad ,, \quad D \; ,, \\ & & & \phi(4, 4) & \quad ,, \quad B \; ,, \end{array}\right\} (\beta).$$

Such terms as (α) or (β) are in order denoted by

$$\left.\begin{array}{llllll} 1s & 2s & 3s & 4s & \text{for the } S \text{ row} \\ & 2p & 3p & 4p & \;\;,, \;\; P \;,, \\ & & 3d & 4d & \;\;,, \;\; D \;,, \\ & & & 4b & \;\;,, \;\; B \;,, \end{array}\right\} \quad \ldots\ldots\ldots(\gamma).$$

Thus the transitions for the various series lines are for

the 'sharp' series, $np \rightarrow 1s$, $n = 2, 3, 4, \ldots,$

the 'principal' series, $ns \rightarrow 2p$, $n = 2, 3, 4, \ldots,$

the 'diffuse' series, $nd \rightarrow 2p$, $n = 3, 4, 5, \ldots,$

the 'Bergmann' series, $nb \rightarrow 3d$, $n = 4, 5, \ldots.$

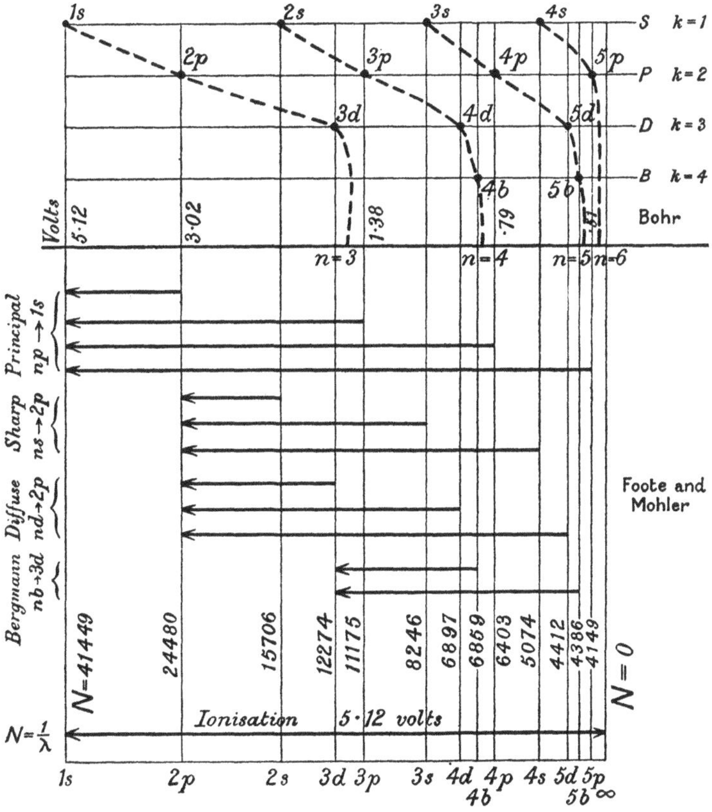

The energy levels of the arc spectrum of sodium.

The diagram on the previous page indicates the energy values of the different terms for sodium, the upper on Bohr's[1] system and the lower on the system of Foote and Mohler[2].

In the diagram the energy of any state (for example the $1s$ state) is expressed both in volts and by a wave number N (the wave number is the number of waves per centimetre and is therefore $1/\lambda$); the line on the extreme right is the level of zero energy, the state when the outer electron is at the periphery of the atom. The number of volts V means the voltage required to drive the outer electron from the given state to the periphery of the atom, so that if W is the negative energy of that state, $W = e\,(V/300)$. The wave number N is the wave number of the radiation which would be emitted by a transition from the boundary of the atom to the given state, and is therefore given by $W = h\nu$, where ν is the frequency. But $\nu = c/\lambda$ and $N = 1/\lambda$. Therefore $\nu = Nc$. Hence $W = hcN$.

Thus for any given state, $eV/300 = hcN$,

$$\therefore\ N = \frac{V \times 4{\cdot}77 \times 10^{-10}}{300 \times 6{\cdot}55 \times 10^{-27} \times 3 \times 10^{10}},$$

or

$$N = 8100V, \text{ approx.}$$

The difference of the N's for any two states is the wave number of the line emitted (or absorbed) in a transition between the two states. Thus if N is found for *any one* state, such as $1s$, it can be found for all the other states from the observed wave numbers of the various spectral lines, which are the *differences* of the N's of the various states concerned.

But the state $1s$ is the 'normal' state of the atom, and it is found that a voltage 5·12 is required to ionise it, that is to drive an electron from the $1s$ orbit to the periphery of the atom. Thus the energy of the $1s$ state is 5·12 volts from which by the approximate equation $N = 8100V$ the corresponding wave number 41472 is found (more accurately 41449).

[1] N. Bohr, 'The Theory of Spectra and Atomic Constitution,' p. 97, Cambridge, 1922.

[2] P. D. Foote and F. L. Mohler, 'The Origin of Spectra,' p. 53, New York, 1922.

85. *Theoretical deduction of the Ritz formula*[1]. For a central field whose potential is $f(r)$, the negative energy W of an n_k orbit was shown (§ 67) to be given by the equation

$$(n-k)\,h = \oint dr \sqrt{-2mW - 2mf(r) - \frac{k^2h^2}{4\pi^2r^2}} \quad \ldots(1).$$

Let a number n_0 be defined by the equation

$$(n_0-k)\,h = \oint dr \sqrt{-2mW + 2m\frac{e^2}{r} - \frac{k^2h^2}{4\pi^2r^2}} \quad \ldots(2),$$

where in the integral $f(r)$ is replaced by $-\dfrac{e^2}{r}$, the Coulomb value. The value of this integral is (p. 64)

$$2\pi\left[\frac{me^2}{\sqrt{2mW}} - \frac{kh}{2\pi}\right],$$

so that

$$n_0h = \frac{2\pi me^2}{\sqrt{2mW}}$$

and

$$W = \frac{2\pi^2me^4}{h^2n_0^2} = \frac{Rch}{n_0^2}.$$

Thus n_0 is the quantum number to be assigned to a Keplerian orbit having the same energy W as the actual orbit in the central field; further, n is not much different from n_0 as for the greater part of its path the electron is effectively under the action of a Coulomb field (§ 82). n_0 is called by many writers the 'effective quantum number' of the orbit.

Subtracting (1) and (2), and noting that the difference of the integrals is a function of k and W, we have

$$(n-n_0)\,h = \phi\,(k, W) = \phi_0\,(k) + \phi_1\,(k).W + \ldots,$$

expanding in terms of W, the ϕ's being small on account of the short time (compared with the whole period) during which the electron is so near the nucleus that the field is non-Coulombian.

[1] G. Wentzel, Zeitschr. für Physik, **19**, p. 53, 1923.

Writing $\phi_0 = h\,\alpha(k), \quad \phi_1 = h\,\beta(k),$

we have approximately

$$n - n_0 = \alpha(k) + W\beta(k),$$

$$\therefore\ W = \frac{Rhc}{n_0^2} = \frac{Rhc}{(n - \alpha_k - W\beta_k)^2},$$

where the suffixes k denote the dependence of α, β on k. This is the Ritz formula.

Bohr has also given a theory of this formula, which not only gives the Ritz formula for the lines of the various series spectra of an element but includes the multiple structure of the lines themselves. Bohr's theory will therefore be given later (§ 90).

CHAPTER XII

THE MULTIPLE STRUCTURE OF SERIES LINES; THEORY OF THE RITZ FORMULA

86. *The multiple structure of the series lines.* In the spectrum of sodium the first line of the principal series, the yellow D line, is found when examined with sufficient power to be really two lines, separated by 6 Å. On the theory so far developed the line is due to the transition $2p \rightarrow 1s$, that is from the orbit ($n = 3$, $k = 2$) to the orbit ($n = 3$, $k = 1$) (see Bohr's diagram, p. 123).

The two lines would be accounted for by supposing that to $n = 3$, $k = 2$ there corresponds not *one* stationary state, but *two*, with slightly different values of the energy. This would require the use of a third quantum number j, so that, the two states corresponding to $n = 3$, $k = 2$ would be $n = 3$, $k = 2, j = 1$ and $n = 3$, $k = 2, j = 2$.

It has been seen in the theory of the Stark and Zeeman effects that the introduction of a definite direction in space (the direction of the electric or magnetic field) leads to a new quantum number connected with the azimuthal angle (ϕ) of the electron round the axis of the field. In like manner, the new quantum number j is to be associated with the axis of the field due to the core (the nucleus and inner electrons), which to this higher degree of approximation is no longer to be regarded as having purely central symmetry.

This doublet separation, in the case of the alkali metals, increases with the atomic number and becomes very pronounced for the heavier metals of the series. For the first line of the principal series the separations are Li (0·25); Na (6·03); K 34·07; Rb (147·35); Cs (422·34) Å.

87. *Theory of the multiple structure.* In the theory of the Zeeman effect it was seen that the effect of the magnetic field could be represented as a precession of an unchanged

Keplerian orbit about the axis of the field; in the theory of the Stark effect, especially in the form given by Bohr (to be given later in § 134), the effect can be represented as a precession of an orbit, of varying form and inclination, about the axis of the electric field.

The theory of the multiple structure which follows supposes that to a first approximation the field may be taken to be central, which leads to a plane periodic motion on which is superposed a rotation in its own plane, as has been shown (chap. VIII). To a higher order of approximation the deviations from the central character have an axial symmetry in space, so that on the above motion is superposed a slow precession round this axis, in a manner analogous to that in the Zeeman and Stark effects.

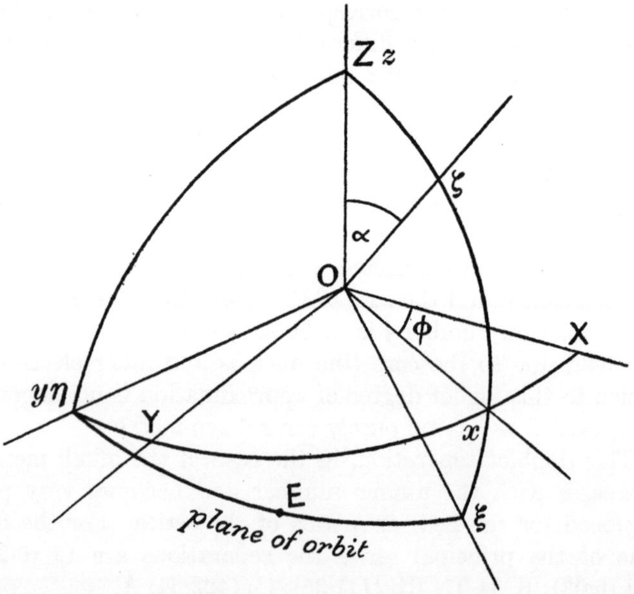

The figure represents three sets of axes OX, OY, OZ; Ox, Oy, Oz; $O\xi$, $O\eta$, $O\zeta$, which meet a unit sphere centre O in the points indicated.

$\zeta = 0$ is the plane of the central field orbit which is precessing about OZ with frequency ω_j; the coordinates of the electron E are $(\xi, \eta, 0)$.

Transforming to the x, y, z axes, we have

$$x = \xi \cos \alpha, \quad y = \eta, \quad z = -\xi \sin \alpha \quad \text{......(1)}.$$

From the theory of the central field (§ 67) we know that ξ, η are of the forms $\Sigma C e^{2\pi i (\tau_n \omega_n \pm \omega_k)}$, where ω_n, ω_k are the frequencies associated with the quantum numbers n, k of that theory.

Hence from (1), x, y, z have the same form, so that we write x, y, z in the form $\Sigma C e^{2\pi i (\tau_n \omega_n \pm \omega_k)}$.

Transforming to the axes X, Y, Z, we have

$$\left.\begin{aligned} X + iY &= e^{i\phi}(x + iy) \\ Z &= z \end{aligned}\right\},$$

and $$\dot{\phi} = \pm 2\pi\omega_j,$$

or $$\phi = \pm 2\pi\omega_j t.$$

Hence X, Y are of the form

$$\Sigma A e^{2\pi i (\tau_n \omega_n \pm \omega_k \pm \omega_j)},$$

and Z is of the form $$\Sigma B e^{2\pi i (\tau_n \omega_n \pm \omega_k)}.$$

By the correspondence principle it follows that while n may change by any value, k can only change by ± 1, and j by ± 1 or 0.

By using these limitations on the changes of the quantum numbers k, j Sommerfeld[1] and Landé[2] have been able to account for the multiple structure of the series spectra in the manner given below.

88. *The doublet systems of the alkali metals*. These are accounted for by orbits for which the values of j are k and $k - 1$. Thus the s, p, d, b states corresponding to $k = 1, 2, 3, 4$ are now resolved into the s; p_1, p_2; d_1, d_2; b_1, b_2 states which correspond to different energy levels for a given n, as shown in the diagram on the next page.

[1] A. Sommerfeld, Ann. d. Phys. **63**, p. 221, 1920.
[2] A. Landé, Zeitschr. für Physik, **5**, p. 231, 1921.

Consider a given line corresponding to a definite n transition. If it is of the principal series, then k changes from 2 to 1; two such transitions are possible, so that the line is a doublet; the switch from a ($k = 2, j = 2$) orbit to a ($k = 1, j = 1$) orbit gives one line of the doublet, and from a ($k = 2, j = 1$) orbit to a ($k = 1, j = 1$) orbit the other.

Thus each line of the principal series when seen under high dispersion becomes a doublet.

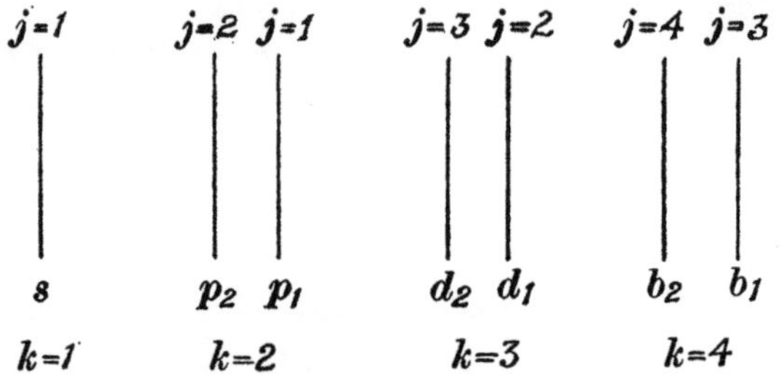

In the case of the D line of sodium, the former of these is the D_2 line ($\lambda = 5896$) and the latter the D_1 line ($\lambda = 5890$). Now the quantum numbers k, j are associated with action variables I_k, I_j such that $I_k = kh$, $I_j = jh$, and I_k, I_j are associated with the coordinates which respectively define the angular motion of the electron round the nucleus and the angular motion (precession) of the orbit round the axis of the field of the inner system (Bohr's theory of § 90). On the analogy that the rotation of a top and its precession tend naturally to be in the same sense, a transition in which k, j change in the same sense is more probable[1] than one where one changes and the other does not; and this latter in turn would be more probable than one where k, j change in opposite senses; thus more atoms would be in the first state of change than the second; and more in the second than the third. Therefore the intensity of the first line would be greater than that of the second, and that of the second than the third.

[1] See N. Bohr, 'Q.L.S.' Part III, p. 104.

Thus the D_2 line for which k, j each decrease by unity should be more intense than the D_1 line for which k diminishes by unity but j does not change; this is a well-known experimental fact.

So the 'sharp' series, depending on transitions from $k = 1$ to $k = 2$, have doublets corresponding to

$$\begin{matrix} k\,1 \\ j\,1 \end{matrix} \rightarrow \begin{matrix} k\,2 \\ j\,1 \end{matrix} \text{ and } \begin{matrix} k\,1 \\ j\,1 \end{matrix} \rightarrow \begin{matrix} k\,2 \\ j\,2 \end{matrix}$$

The latter are more intense than the former.

For the 'diffuse' series where switches occur from $k = 3$ to $k = 2$, the possibilities are

$$\begin{matrix} k = 3 \\ j = 3, 2 \end{matrix} \text{ to } \begin{matrix} k = 2 \\ j = 2, 1 \end{matrix}$$

and since j can only change by $\pm$ 1 or 0, the actual transitions are

$$\begin{matrix} k\,3 \\ j\,3 \end{matrix} \rightarrow \begin{matrix} k\,2 \\ j\,2 \end{matrix}; \quad \begin{matrix} k\,3 \\ j\,2 \end{matrix} \rightarrow \begin{matrix} k\,2 \\ j\,2 \end{matrix}; \quad \begin{matrix} k\,3 \\ j\,2 \end{matrix} \text{ to } \begin{matrix} k\,2 \\ j\,1 \end{matrix}$$

The first and third are strong lines, but the second is much fainter (a so-called 'satellite'). Thus each line resolves into a doublet, with a faint satellite, under high dispersion.

89. *The triplet systems of the alkaline earths.* An illustration

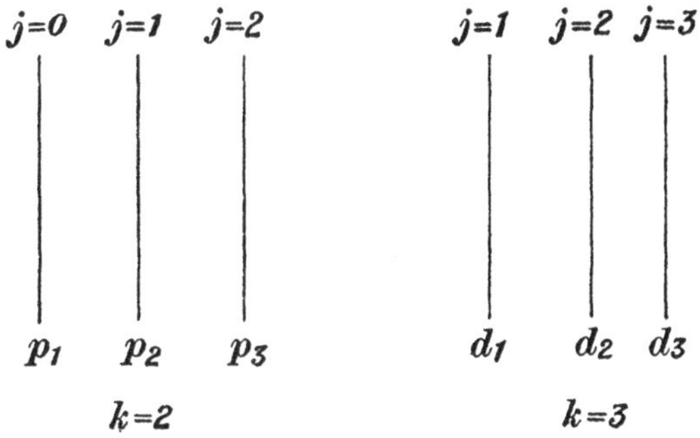

will suffice: the case of the diffuse series where k changes from 3 to 2. For each k, we assume three values of j, viz. k, $k - 1$, $k - 2$ giving three energy levels for each value of k.

There are nine possible transitions:

$$\left.\begin{matrix} k\,3 \\ j\,3 \end{matrix} \rightarrow \begin{matrix} k\,2 \\ j\,0,1,2 \end{matrix}\right\} \cdot \left.\begin{matrix} k\,3 \\ j\,2 \end{matrix} \rightarrow \begin{matrix} k\,2 \\ j\,0,1,2 \end{matrix}\right\} \quad \left.\begin{matrix} k\,3 \\ j\,1 \end{matrix} \rightarrow \begin{matrix} k\,2 \\ j\,0,1,2 \end{matrix}\right\}.$$

Since j can only change by $\pm$ 1 or 0, $j = 0$, 1 cannot occur in the first bracket, nor $j = 0$ in the second.

Thus the actual switches are $j3 \rightarrow j2$; $j2$ to $j1$; $j2$ to $j2$; $j1$ to $j0$; $j1$ to $j1$; $j1$ to $j2$, and in each $k3 \rightarrow k2$.

Thus the first, second and fourth are strong lines, forming a 'triplet,' the third and fifth are fainter (primary satellites), and the sixth fainter still (secondary satellite); so that each line of the diffuse series appears as a triplet, with three satellites, two primary and one secondary.

90. *Bohr's*[1] *generalisation of the Ritz formula.* The motion, as in § 87, is taken to be of the rosette type in a plane precessing round the axis of the atomic field. As seen in that paragraph, the motion is a triply periodic one with frequencies ω_n, ω_k, ω_j. The stationary states are therefore specified by $I_n = nh$, $I_k = kh$, $I_j = jh$ and in transitions k can change by $\pm$ 1 and j by $\pm$ 1 or 0.

The I's satisfy the usual condition

$$\delta E = \omega_n \delta I_n + \omega_k \delta I_k + \omega_j \delta I_j \qquad (\S\ 53),$$

which refers to two states of the atom where the inner electrons are unchanged and the outer electron changes its path slightly.

Since the perturbations of the orbit from a Keplerian ellipse (§ 82) occur almost entirely close to perihelion, where the electron is for only a small part of a period, this period will to a high approximation be that of the Kepler ellipse of which the outer loop forms a part.

With this approximation, $\delta E = \omega_n \delta I_0$, where I_0 is the value of I_n for this ellipse, which is connected with ω_n by the equation

$$\omega_n = \frac{4\pi^2 m e^4}{I_0^{\,3}} \qquad (\text{p. } 47).$$

$$\therefore\ \omega_n \delta I_0 = \omega_n \delta I_n + \omega_k \delta I_k + \omega_j \delta I_j.$$

[1] N. Bohr, Proc. Phys. Soc. London, 35, p. 296, 1923.

Hence we may write $I_0 = I_n + \Phi(I_k, I_j)$, where Φ is a function of I_k, I_j, subject to the relation

$$\omega_n \delta\Phi = \omega_k \delta I_k + \omega_j \delta I_j,$$

from which it follows that ω_k/ω_n, ω_j/ω_n, with the above approximation, are independent of I_n.

From $$\delta E = \omega_n \delta I_0 = \frac{4\pi^2 m e^4}{I_0^3}\delta I_0,$$

we have $$E = -\frac{2\pi^2 m e^4}{I_0^2},$$

as in § 32.

$$\therefore\ W = -E = \frac{2\pi^2 m e^4}{I_0^2} = \frac{2\pi^2 m e^4}{\{I_n + \Phi(I_k, I_j)\}^2},$$

and inserting the values of the I's,

$$W = \frac{2\pi^2 m e^4}{h^2}\cdot\frac{1}{\{n + \psi(k,j)\}^2} = \frac{Rhc}{[n + \psi(k,j)]^2}.$$

This is Bohr's result and is the Ritz formula in a generalised form, which accounts not only for the series lines themselves, but also for their multiple structure.

91. *The mechanical significance of Ritz's formula (Bohr)*[1]. The energy 'term' of the hydrogen atom with its Keplerian orbit is $E = \frac{-Rhc}{n^2}$. The frequency

$$\omega_0 = \frac{\partial E}{\partial I} = \frac{\partial E}{h\partial n} = \frac{2Rc}{n^3} = 2Rc\left(\frac{-E}{Rhc}\right)^{\frac{3}{2}} = f(E) \text{ suppose.}$$

The Rydberg formula $E = \frac{-Rhc}{(n-\alpha)^2}$ gives a frequency

$$\omega = \frac{\partial E}{h\partial n} = \frac{2Rc}{(h-\alpha)^3} = 2Rc\left(\frac{-E}{Rhc}\right)^{\frac{3}{2}} = f(E),$$

as before.

Thus for a given E, $\omega = \omega_0$.

The Ritz formula

$$E = \frac{-Rhc}{(n - \alpha - \beta E)^2}$$

[1] A note given to the author in a recent conversation with Professor Bohr in Copenhagen, 1925.

leads to a frequency

$$\omega = \frac{\partial E}{h\partial n} = \frac{2Rc}{(n-\alpha-\beta E)^3}\left(1-\beta\frac{\partial E}{\partial n}\right)$$

$$= 2Rc\left(\frac{-E}{Rhc}\right)^{\frac{3}{2}}(1-h\beta\omega)$$

$$= f(E)\,(1-h\beta\omega).$$

Thus for a given E

$$\omega = \omega_0\,(1-h\beta\omega),$$

$$\therefore\ \frac{1}{\omega}-\frac{1}{\omega_0} = h\beta.$$

Therefore if σ, σ_0 are the periods,

$$\sigma - \sigma_0 = h\beta.$$

Thus the β term in Ritz's formula is a measure of the deviation of the period from that in a Keplerian orbit of the same energy; the small β term indicates that for the greater part of the time the path of the electron is a Keplerian orbit, the time spent in penetrating the inner system being a small part of the period.

CHAPTER XIII

ARC AND SPARK SPECTRA; EXCITATION OF LINES BY ELECTRONS OR RADIATION

92. *Arc and spark spectra.* If the electric arc is used to excite a spectrum, one electron is driven to an outer orbit, and by its transitions to other orbits of less energy causes the emission of the '*arc spectrum.*'

If the condensed spark discharge is used, one or more electrons are expelled from the atom altogether (so that it is ionised) and one of the remaining electrons, driven to an outer orbit, causes the emission of a '*spark spectrum.*'

Take a definite case, that of Mg (atomic number 12); the arc spectrum is due to an outer electron which moves under the action of the nucleus ($12e$) shielded by 11 electrons; at large distances the electron is effectively moving under the action of a charge e, as in the hydrogen atom. Hence the series formula must become the same as the hydrogen formula for large quantum numbers; for example the Rydberg formula $W = Rhc/(n - \alpha_k)^2$ becomes the hydrogen formula $W = Rhc/n^2$ for large values of n.

The spark spectrum, with *one* electron ejected from the atom, is due to an outer electron moving under the action of the nucleus ($12e$) shielded by 10 electrons; at large distances the electron is effectively moving under the action of a charge $2e$, as in the case of ionised helium. The series formula must therefore become $4Rhc/n^2$ for large values of n, and the Rydberg formula becomes

$$W = 4Rhc/(n - \alpha_k')^2.$$

If *two* electrons are ejected, the outer electron at large distances moves under the action of an effective charge $3e$, as in Li_{++}, and $W = 9Rhc/(n - \alpha_k'')^2$.

These three spectra, the arc spectrum, the spark spectrum with one electron ejected from the atom, and the spark spectrum with two electrons ejected will be referred to as simply the spectra of Mg, Mg_+, Mg_{++}.

In the table of the elements, Na, Mg, Al, Si occur in succession with atomic numbers 11, 12, 13, 14. Therefore the spectra of Na, Mg_+, Al_{++}, Si_{+++} are due to an outer electron revolving round a core consisting of the nucleus and 10 electrons, the nuclear charge being $11e$, $12e$, $13e$, $14e$ in the four cases. This only affects the Rydberg constant, which would be R, $4R$, $9R$, $16R$ in the corresponding formulae; otherwise the spectra are the same, as the shielding effect of the 10 electrons is the same in all four cases.

Again, the alkali metals are preceded in each case in the table of atomic numbers by an inert gas; sodium is preceded by neon; potassium by argon; rubidium by krypton; caesium by xenon.

Hence if I denotes an inert gas and A the corresponding alkali, the spectrum of A_+ should on the theory be similar to that of I.

This is in striking agreement with experiment, for while the arc spectra of the alkalis have a comparatively simple structure, the spark spectra ('enhanced' spectra) of the alkalis are extremely complex in structure, like those of the inert gases.

93. *The building of atoms.* Bohr has often in his writings pointed to the importance of spectra as representing the process of building up atoms step by step. The hydrogen spectrum represents the process of binding the first electron of an atom to the nucleus. The strength of the binding is measured by the energy required to dislodge it from the 'ground' orbit and send it to the edge of the atom. For hydrogen this is 13·54 volts.

In the case of the *arc* spectrum of Mg (atomic number 12), for example, the difference of wave number of the ground orbit of the 'principal' series and of the limit of the series can be determined, and expressed in volts (§ 84) is the work required to drive the 12th electron from the ground orbit to the edge of the atom; it is therefore the strength of binding of the 12th electron and is found to be 7·61 volts. The *spark* spectrum, i.e. the spectrum of Mg_+, gives corresponding evidence as to the binding of the 11th electron, the 12th being out of the picture (expelled from the atom) and the strength of binding of the 11th electron is found to be 7·36 volts.

94. *Excitation of spectral lines by electron bombardment.* The excitation of spectral lines in a gas or vapour by bombardment of the atom with a stream of electrons was first carried out by Franck and Hertz[1] in 1914. Electrons with gradually increasing velocities (produced by increasing the voltage through which they fall) were projected into a mass of mercury vapour; this vapour, being monatonic, the results are not complicated by the dissociation of molecules. At first the collisions are 'elastic,' the electron rebounding from the atom without loss of energy, but when a certain critical velocity was reached, corresponding to 4·86 volts, the collisions became 'inelastic' and the energy of the electron was completely absorbed by the atom; at the same instant the ultra-violet line $\lambda = 2537$ of mercury was emitted, and thus a *one line* spectrum produced.

On Bohr's theory, the absorption of the energy has driven the active outer electron from its orbit in the 'normal' state to the next stationary state of higher energy (or lower W); the spontaneous return from this state to the normal causes the emission of the spectral line. On this view the electron, at the critical velocity, must have possessed kinetic energy exactly equal to the one quantum $h\nu$ required for the transition.

Hence $h\nu = \frac{1}{2}mv^2 = e\,(V/300)$, where V is the voltage producing v; or if λ is the wave length,

$$\frac{hc}{\lambda} = \frac{eV}{300} \text{ or } \lambda V = \frac{300hc}{e}.$$

[1] J. Franck and G. Hertz, Verh. d. D. Phys. Ges. **16**, p. 512, 1914 et seq.; Phys. Zeitschr. **20**, p. 132, 1919.

Using the known values of h, c, e, this equation is

$$\lambda = \frac{12345}{V} . 10^{-8} \text{ cm.},$$

or
$$\lambda = \frac{12345}{V} \text{ Ångström units.}$$

Thus on this view, $\lambda = 2537$ Å. would correspond to

$$V = \frac{12345}{2537} = 4{\cdot}9 \text{ volts},$$

which is in close agreement with experiment.

The next line $\lambda = 1849$ Å. is excited at a voltage of 6·7 which again is given by the equation of the theory. At voltages between 4·9 and 6·7 the first line continues to be excited alone, the atom taking the necessary quantum from the electron, which goes on with diminished kinetic energy.

Thus line after line can be excited, each new one at a definite value of the voltage controlling the electron; these voltages are called the *resonance potentials* of the different lines (Tate and Foote). The voltage necessary to excite the whole series of lines up to its head, so that the active outer electron of the atom is just driven to the periphery, is called the *ionisation potential*, and for mercury vapour is 10·4 volts.

These results are a convincing confirmation of Bohr's theory of spectral emission. On the classical theory the lines must be emitted simultaneously or not at all, being due to different harmonic components of *one* motion, whereas in Bohr's theory the emission of each line is a separate process in accord with emission of the lines one after another observed by Franck and Hertz.

95. *Excitation of spectral lines by light of given frequency—Resonance radiation.* Füchtbauer[1] first showed that the excitation of a spectral line can be effected by using light of that frequency instead of an electron.

Wood and Dunoyer[2] found that if sodium vapour is subjected to radiation from *one* of the two components of the

[1] C. Füchtbauer, Phys. Zeitschr. **21**, p. 635, 1920.
[2] R. W. Wood and L. Dunoyer, Phil. Mag. **27**, p. 1018, 1914.

yellow line, say the D_2 line ($\lambda = 5896$), the resonance radiation, at least at low pressures, consists only of this component. The exciting radiation drives the electron from the orbit $1s$ to the orbit $2p_2$ and the spontaneous return from $2p_2$ to $1s$ gives the resonance line D_2.

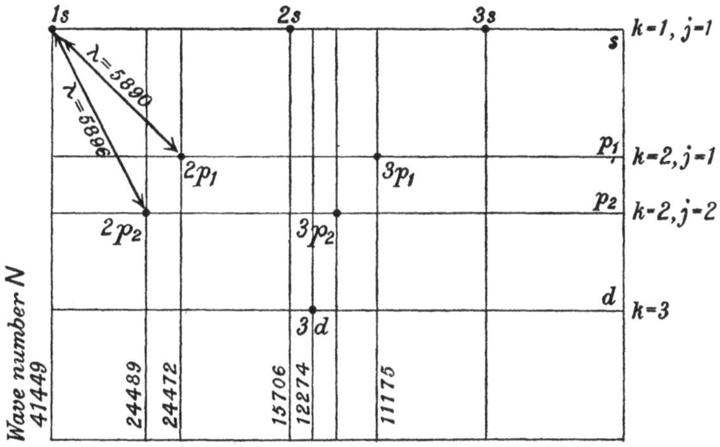

The doublets of the principal series of sodium.

But there is a fundamental difference between excitation of lines by incident radiation and by electrons; in the case just referred to, the energy absorbed by the atom is the quantum $h\nu_0$ required to drive the atomic electron from the $1s$ to the $2p_2$ orbit, where ν_0 is the frequency of the radiation corresponding to $2p_2 \rightarrow 1s$.

If an incident electron has *any* velocity, with energy $\geqslant h\nu_0$, it can part with the necessary quantum and go on at reduced speed; but incident radiation can only part with energy to the atom in its *own* quanta determined by its own frequency ν_1, so that ν_1 must be *exactly* equal to ν_0 if the change in the atom necessary for the emission is to be effected.

These experiments on sodium vapour were continued by Strutt[1], who used as the exciting radiation that of the second line of the principal series $3p \rightarrow 1s$, $\lambda = 3303$; the resonance

[1] R. J. Strutt, Proc. Roy. Soc. **96**, p. 272, 1919.

radiation excited however consisted only to a very small extent of light of this frequency, the main part being that of the ordinary yellow *D* line (the first line of the principal series). This result is quite contrary to the ideas of resonance in the ordinary dynamical theory, as there is no rational connection between the frequencies of the first and second lines of the principal series. But this curious result meets with an immediate explanation on Bohr's theory.

If we take the exciting radiation to be that of the line $3p_1 \rightarrow 1s$, one of the doublets of the second line of the principal series, the electron is driven from the normal $1s$ orbit to the $3p_1$ orbit. There are many ways in which it can return to $1s$; the possibilities are

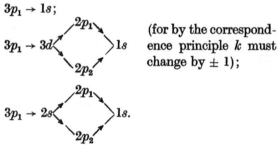

$3p_1 \rightarrow 1s$;

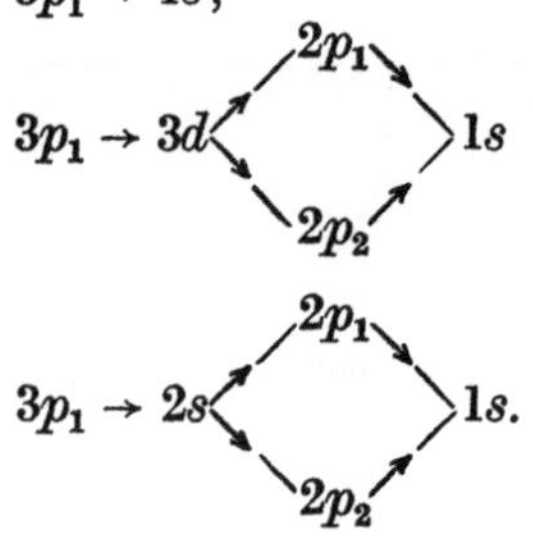

(for by the correspondence principle k must change by $\pm$ 1);

From the diagram, by subtraction of the corresponding wave numbers, these five processes lead to the emission of lines of the following wave numbers:

(i)		30274;	
(ii)	1099,	12198,	16977;
(iii)	1099,	12215,	16960;
(iv)	4531,	8766,	16977;
(v)	4531,	8783,	16960.

All these lines except 30274 and $\begin{matrix}16960\\16977\end{matrix}$ are in the infra-red and would not be observed in these experiments; the former is the second line and the latter pair are the components of the first line (the *D* line) of the principal series; the experiments show that the last four forms of transition occur in many more atoms than the first does, as the emitted radiation is almost wholly that of the D line.

The theory completely explains how the exciting radiation $\lambda = 3303$ causes the emission of the yellow D lines $\lambda = 5890$, 5896. Also *both* the components of the D doublet are excited though light from only *one* component of the original doublet was used to excite them.

96. *Optical spectra and the periodic law.* The changes which give rise to optical spectra occur in the outer group of electrons. The alkali metals all have 8 electrons in the outermost complete group and one electron beginning a new group outside them; on account of the similarity of the outer grouping the alkalis all have similar spectra; the increase of atomic number does not show itself in the series types, but only in a secondary manner in the multiple structure of the lines of a series (where the separation increases in magnitude with atomic number (§ 86)).

So the alkaline earths have similar spectra, each having two electrons outside the next complete group of 8 electrons.

The periodicity of structure of the atom is indicated by the periodicity in the character of the spectrum.

CHAPTER XIV

X-RAY SPECTRA

97. *X-ray spectra.* When kathode rays impinge on an anti-kathode coated with a given substance, the X-rays emitted, when resolved into a spectrum by a crystal[1], are found to have two spectra, one a *continuous* spectrum whose upper frequency corresponds to the velocity of the kathode rays (§ 25), and the other a *line* spectrum characteristic of the substance. In the latter, every element shows the same grouping of the lines, but with increasing atomic number the groups move towards regions of shorter wave length. These line spectra each contain several series of lines, the $K, L, M, \ldots$ series, which are quite distinct from one another, and do not overlap as do the series of the optical spectra.

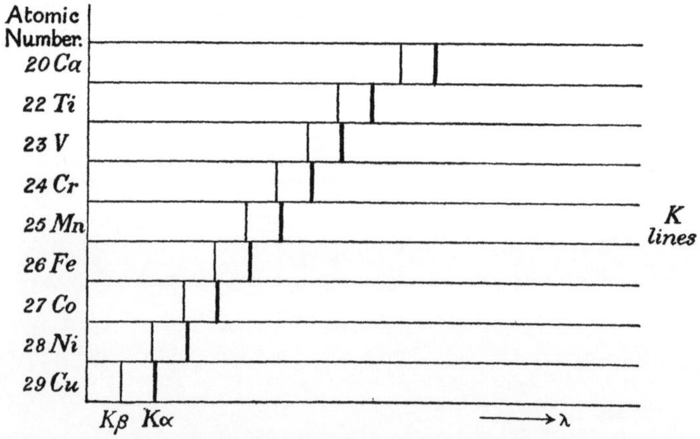

In 1913, Moseley[2] photographed the K series of lines for the elements calcium to copper, and the figure shows the

[1] W. H. and W. L. BRAGG, 'X-rays and Crystal Structure,' 1915.
[2] H. G. J. MOSELEY, Phil. Mag. **26**, p. 1024, 1913; **27**, p. 703, 1914.

decrease of wave length with increasing atomic number Z. Though the order of the elements iron, cobalt, nickel in the table of atomic weights is Fe, Ni, Co, the atomic weights being 55·84, 58·68, 58·97, the X-ray spectrum shows that in atomic number the order should be Fe, Co, Ni. (Cobalt is a mixture of isotopes of cobalt of atomic weights 58, 60 and nickel is a mixture of isotopes of nickel also of atomic weights 58, 60.)

In Moseley's figure, the now known rare element Sc (atomic number 21) is missing between Ca and Ti; its absence is clearly shown by a gap in the regular sequence. Other gaps for undiscovered elements are shown by X-ray spectra to correspond to $Z = 43, 61, 75, 85, 87$.

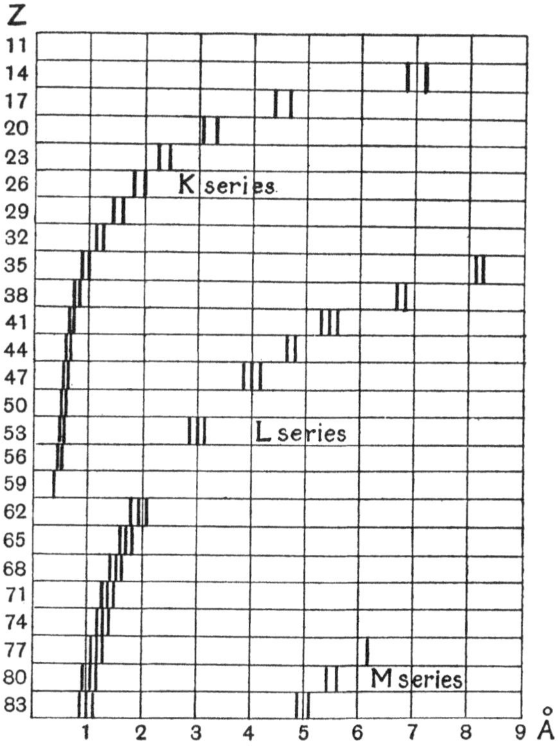

The diagram on p. 143, due to Siegbahn[1], shows the nature of the K, L, M series for a large range of atomic number, and the range of elements for which the various series could be observed.

98. *Kossel's*[2] *theory of X-ray spectra.* In order to produce a K line, the atom must have an electron ejected from the K shell to beyond the shells occupied by other electrons, that is (so far as the voltage necessary is concerned) effectively to the periphery of the atom; its place can be taken by the transition of an electron from one of the L, M, ... shells, with a corresponding emission of radiation. A switch from $L \to K$ produces the first line of the K series, the K_α line; one from $M \to K$ produces the K_β line, and so on.

So gaps in the L shell can be filled up by transitions from the M, N, ... shells giving rise to the L series of lines, and so on.

Thus the emission of an X-ray spectrum is due to the restoration of the stable state of the inner electrons after one of their number has been driven out by the exciting cause of the spectrum. If the disturbing cause is only sufficient to wrench out an electron from the L shell, then only L, M, ... lines are produced.

The K lines tend to a limit, which corresponds to the fall of an electron from the edge of the atom to a K orbit. This limit it will be seen is found from absorption spectra (§ 104), and if the frequency is ν, $h\nu$ is the energy associated with the K group.

For uranium (atomic number 92), the corresponding frequency expressed in volts by $h\nu = eV/300$, is 115,000 volts[3], which is the voltage necessary to excite the K lines. For a light atom such as magnesium (atomic number 12) it is only 1300 volts.

[1] M. Siegbahn, Verh. d. D. Phys. Ges. **18**, p. 278, 1916; Jahrbuch Rad. u. Elek. **13**, p. 296, 1916.

[2] W. Kossel, Verh. d. D. Phys. Ges. **16**, pp. 899, 953, 1914; **18**, p. 339, 1916.

[3] W. Duane, Proc. Nat. Acad. **6**, p. 607, 1920.

99. *The relation of wave number to atomic number.* Moseley plotted $\sqrt{\frac{N}{R}}$, where N is the wave number and R the Rydberg constant, against the atomic number Z as abscissa for the K_α lines of the different elements. He found that the curve was a straight line, which when produced to meet the axis of zero wave number cut it in the point $Z = 1$. The slope of the line was about ·866 or $\sqrt{\frac{3}{4}}$.

Therefore $\sqrt{\frac{N}{R}} = \sqrt{\frac{3}{4}}(Z-1)$ expresses Moseley's experimental result.

$$\therefore \quad N = R(Z-1)^2\left(\frac{3}{4}\right)$$

$$= R(Z-1)^2\left(\frac{1}{1^2} - \frac{1}{2^2}\right).$$

Thus the wave numbers are the same as if an electron revolved alone about a nucleus $(Z-1)e$ and a transition occurred from an orbit of quantum number 2 to one of quantum number 1, just as in the first line of the Lyman series for hydrogen. This means that the nuclear charge, diminished in effective value by e owing to the presence of the negative electrons near it, controls the orbits of the L and K levels concerned with the emission of the K_α line, the effect of the outer structure of the atom being negligible. As it is to this outer structure that the periodicity of the chemical and optical properties is due, it is apparent why the K lines give no indication of such periodicity but only indicate growing atomic number.

The L lines give similar results, though in a less clear-cut form as the control of the nucleus is less direct than for the K lines, owing to the intervening electrons.

The $M, N, \ldots$ lines are due to transitions closer and closer to the outermost electrons of the atom, so that as we proceed along the series the correspondence with optical spectra must begin to show itself.

100. *Excitation of the X-ray spectrum.* The figure indicates the energy levels of a heavy element, caesium (atomic number 55). In the K, L, M, N, O, P groups of orbits or 'shells' there are respectively 2, 8, 18, 18, 8, 1 electrons; the dotted lines indicate energy levels of unoccupied orbits up to the edge of the atom shown by the dotted line (∞), which is, as usual, the level of zero energy. Some of the energy values are indicated. In order to excite the K lines, the voltage must be enough to drive an electron from the K level to at least the P level, which is the first level not completely

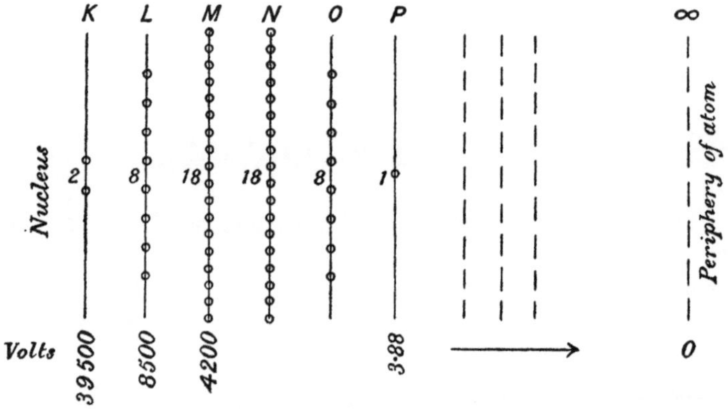

occupied. This voltage is seen to be (39,500 – 3·88) or effectively 39,500, which would drive it to the periphery of the atom. Transitions in the different atoms from the $L, M, \ldots$ levels to fill up the gap then cause the emission of the K_α, K_β lines.

It is clear that no K line will be excited until the voltage reaches 39,500, when all the K lines will be excited simultaneously, in contrast to the separate excitation of the optical lines (§ 94). So all the L lines are excited at 8500 volts and the M lines at 4200.

101. *The connection between optical and X-ray spectra.* This, as has been indicated, must be sought for in the lighter elements, where on account of the fewness of the 'shells,' the

X-ray transitions cannot be far from the surface 'shell' which is concerned with the optical spectra.

Consider the element sodium (atomic number 11) and suppose the figure to indicate energy levels just as for caesium, above.

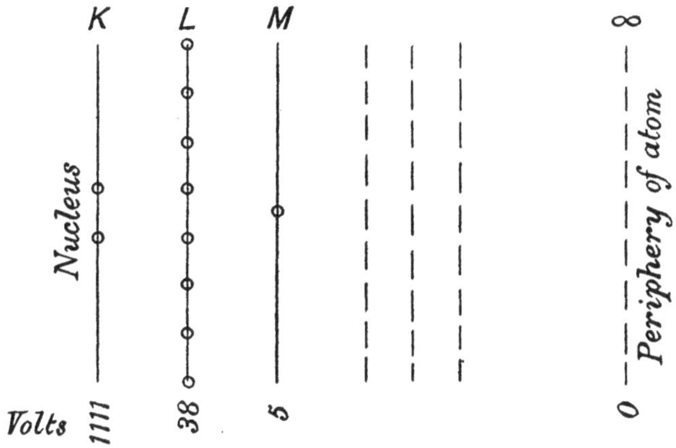

Here (1111 − 5) or 1106 volts are needed to drive an electron from the K level to an unoccupied orbit, so that the filling up of the gap by transitions from L, M to K would excite the K_α, K_β lines.

A voltage (38 − 5), or 33, would drive an electron from the L to the M level and by a transition of it or the one already there to the L level, the L_α line is emitted. The energy emitted in this transition, 33 volts, is equivalent (from the formula $V = 12345/\lambda$ of § 94) to a wave length of 374 Å.; so that this L_α line corresponds to an X-ray of $\lambda = 374$ Å., which is a very short ultra-violet ray. As such it has actually been observed by Millikan[1] by an extension of the range of the diffraction grating; he found a wave length 372·2 Å.

At neon (atomic number 10), the L series must cease as there are no electrons to fall into the L shell when an electron is dislodged from it. So the K series terminates with helium, and the M series with argon.

[1] R. A. Millikan, Proc. Nat. Acad. Sci. 7, p. 289, 1921.

CHAPTER XV

ABSORPTION SPECTRA AND ABSORPTION EDGES

102. *Absorption spectra.* On the classical theory of radiation all the emission lines of a vapour should appear as absorption lines when a column of vapour is viewed against a source emitting a continuous spectrum.

On Bohr's theory, to take a definite case, that of sodium, the 'outer' electron in the normal unexcited atom is moving in a $1s$ orbit. This atom can absorb radiation whose quantum $h\nu$ will exactly suffice to cause a transition from $1s \rightarrow np$ (since k must change only by ± 1), that is, any radiation whose frequency is the same as that emitted by a transition $np \rightarrow 1s$. Thus the absorption spectrum consists of lines corresponding to the principal series only, in general.

When the head of the series is reached, corresponding to frequency ν_H, the quantum $h\nu_H$ absorbed is just sufficient to drive the electron from $1s$ to the periphery of the atom; radiation of frequency ν greater than ν_H is completely absorbed for all values of ν, as the part of $h\nu$ over and above that required to drive the electron to the edge of the atom is used in endowing it with kinetic energy $\frac{1}{2}mv^2$ equal to $h\nu - h\nu_H$. Outside the atom the electron may have any value for the energy and not necessarily those of certain stationary states as it must have inside. Thus beyond the head of the series there is a continuous band of absorption.

Therefore the absorption spectrum consists of absorption lines coincident with the principal series lines and a continuous absorption band starting from the head of the series. R. W. Wood[1] has observed some 50 absorption lines corresponding to the principal series for sodium and the continuous band from the head of the series to the extreme ultra-violet.

[1] R. W. Wood, 'Physical Optics,' p. 513, 1911.

103. *Absorption spectra of excited atoms.* Atoms excited by incident radiation or even by a sufficient elevation of temperature may have the 'outer' electron in an orbit other than the normal one. If it is at the next higher energy level, for example, at the level $n = 2$ in the case of hydrogen, then the Balmer lines can appear as absorption lines; in the case of the alkalis with the electron in a $2p$ orbit instead of the normal $1s$ orbit, the 'sharp' or 'diffuse' series might appear as absorption lines.

The spectrum of ζ Tauri shows the Balmer absorption lines from H_∞ to H_{31}; they are due to the absorption by a layer of 'excited' H atoms of the continuous emission spectrum from the inner layers of the star below.

Foote and Mohler[1] have given a full discussion of the effect of temperature on absorption spectra, and illustrated its importance as an indication of stellar temperatures.

104. *X-ray absorption edges.* If X-rays, spread out into a continuous spectrum by a revolving crystal, are passed through a thin plate containing the element to be examined, an absorption spectrum can be taken.

In the case of a heavy element such as caesium, the frequencies of the K, L, M, ... lines are high, and 39,500 volts are needed to excite the K lines, 8500 the L lines, 4200 the M lines and so on. But only about 4 volts are needed to excite the whole optical spectrum, that is to drive the single electron from the P shell to the edge of the atom (see the figure of p. 146); hence the various optical energy levels may be neglected in considering X-ray phenomena as their range is only 4 volts.

Rays from the incident beam of frequency corresponding to 8500 volts are fully absorbed by the atom of caesium, as this voltage is just enough to expel an L electron to the first unoccupied orbit in the P shell or what is effectively in this case the edge of the atom, and so excite the whole of the

[1] P. D. Foote and F. L. Mohler, 'The Origin of Spectra,' chap. VII, New York, 1922.

L series. Frequencies greater than this are also fully absorbed, for the excess of the voltage over 8500 gives kinetic energy to the electron after its expulsion from the atom. These electrons by falling back into the atom give rise to fluorescent radiation, increasing with the kinetic energy given to the electron but scattered, so that the effective absorption becomes less and less with increasing voltage, until at 39,500 volts the *K* series is excited and there is again a strong absorption which gradually diminishes with increase of the voltage.

Thus there is a series of absorption bands terminating sharply at the *K*, *L*, *M*, ... excitation frequencies; the edges of these bands are the *K*, *L*, *M*, ... 'absorption edges.'

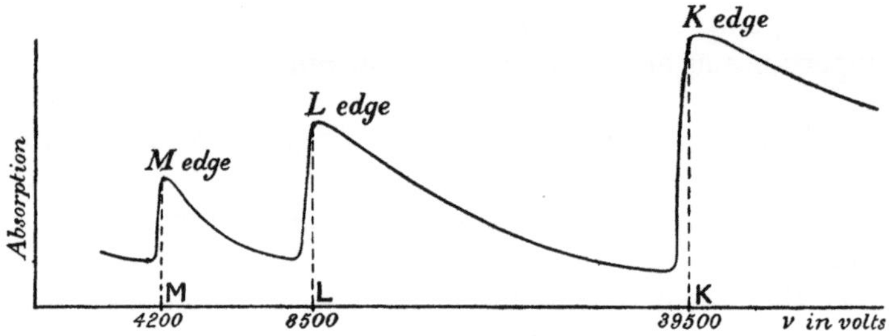

This is indicated in the figure; in a photograph the band *LK* would be light at *L* and would increase in darkness up to *K*, lightness in the photograph corresponding to a large absorption of the incident rays by the substance.

But for a light element, such as sodium (atomic number 11), where the voltages in the optical region are no longer negligible compared with those in the X-ray region, the *K*, *L*, *M* edges lose their sharpness.

At 5 volts all the optical absorption lines are in evidence and there is full absorption at the *M* absorption edge. As the voltage increases the absorption decreases (owing to fluorescence) until the voltage 33 (38 − 5) is reached, which suffices to drive an *L* electron to an *M* orbit and excite the first *L* line; as the voltage rises from 33 to 38, *L* electrons are driven to various optical levels and so excite other L lines, so that there is 'selective' absorption superposed on the general decrease of absorption due to fluorescence, which reaches a

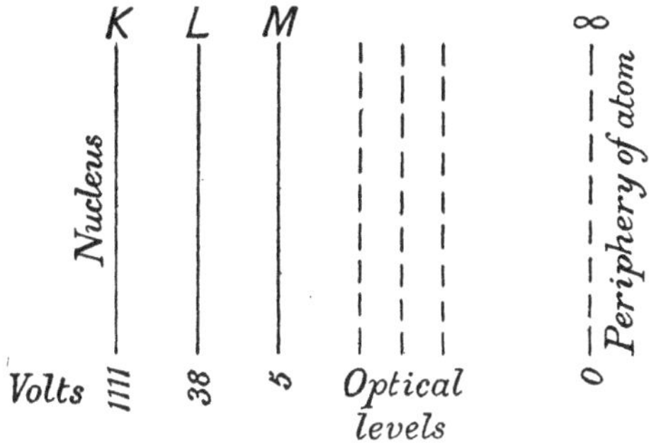

maximum at 38 volts; then there is a decrease, until at 1106 volts (1111 – 5) a K electron is driven to the M level, and there is again the same kind of selective absorption between 1106 and 1111 volts, when there is a maximum at 1111 volts, followed by a decrease as before.

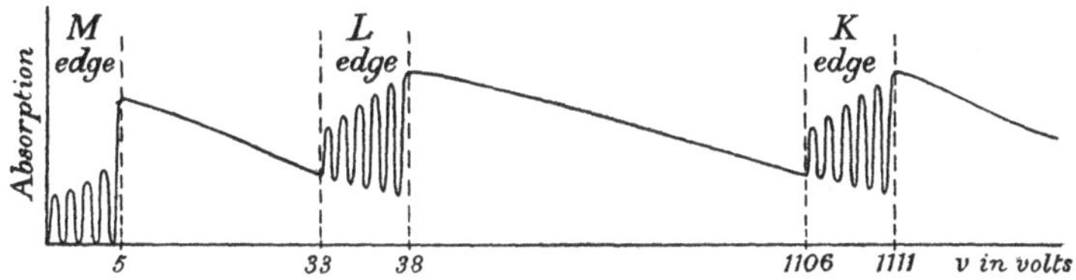

Thus on this theory, at each edge there is a region of selective absorption of a range of about 5 volts. This structure of the edges has been observed by Stenström[1] for the M edge, Hertz[2] for the L edge and Fricke[3] for the K edge, for some of the lightest elements.

105. *X-ray levels of the atom.* The absorption edges determine the energy values of the orbits of the K, L, M groups in the atom. For the K series, one edge is observed; for the

[1] W. Stenström, Dissertation, Lund, 1919.
[2] G. Hertz, Zeitschr. für Physik, **3**, p. 19, 1920.
[3] H. Fricke, Phys. Rev. **16**, p. 202, 1920.

L series two or three; for the *M* series three or four or five; these results are mainly due to the exact measurements of Siegbahn[1] and his collaborators at Lund.

The diagram, due to Bohr and Coster[2], shows the nature of the *K*, *L*, *M* limits for the different elements, the abscissa being the atomic number.

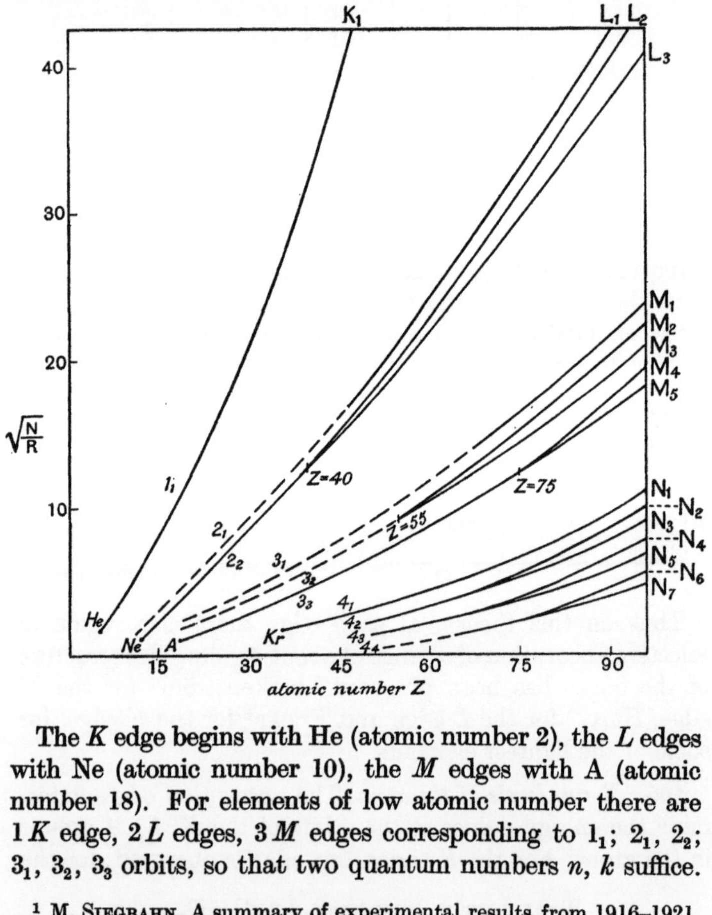

The *K* edge begins with He (atomic number 2), the *L* edges with Ne (atomic number 10), the *M* edges with A (atomic number 18). For elements of low atomic number there are 1 *K* edge, 2 *L* edges, 3 *M* edges corresponding to 1_1; 2_1, 2_2; 3_1, 3_2, 3_3 orbits, so that two quantum numbers *n*, *k* suffice.

[1] M. Siegbahn, A summary of experimental results from 1916–1921, Jahrb. d. Radioaktivät, **18**, p. 240, 1921.

[2] N. Bohr and D. Coster, Zeitschr. für Physik, **12**, p. 342, 1923.

But one L level splits into two at about $Z = 40$ and two of the M levels each split into two, one at $Z = 55$ and the other at $Z = 75$. This indicates that the stationary states at these points require a new quantum number characterising the change of the atom from its normal state to a new one which appears after the removal of one of the inner electrons (Bohr)[1].

Coster and also Wentzel have constructed a diagram of energy levels for niton (atomic number 86) based upon measurements of the spectra of heavy elements in the neigh-

Niton

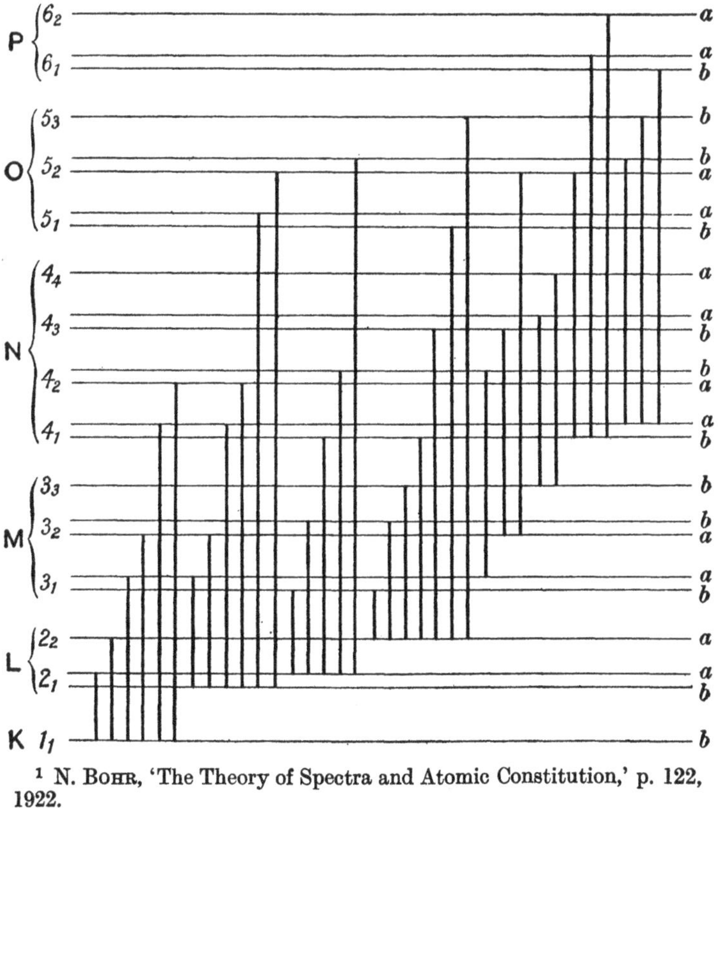

[1] N. BOHR, 'The Theory of Spectra and Atomic Constitution,' p. 122, 1922.

bourhood of niton. This gas has electrons in the *K*, *L*, *M*, *N*, *O*, *P* shells, on Bohr's theory (chap. XVI), the sixth shell having its full number of electrons, 8. The conclusions of these writers agree with Bohr's general scheme, which gives niton orbits up to quantum number 6.

The various n_k orbits are shown and the horizontal lines represent energy values in order of magnitude. The lines joining pairs of levels indicate transitions corresponding to observed lines.

Corresponding to given values of *n*, *k* there are in most cases two energy values, i.e. two orbits, denoted by *a* and *b*, corresponding to two values of some third quantum number. It was found by Coster and Wentzel that transitions are limited by two conditions, (i) that *k* changes by ± 1 or 0, (ii) that they must be between an *a* level and a *b* level. A discussion of the outlying difficulties of this part of the subject has been given by Bohr[1], and in a recent paper Coster[2] has extended the classification of levels to the lighter elements.

106. *Photo-electric determination of absorption edges.* De Broglie[3] has found absorption limits by measuring the velocities of electrons ejected from the atom by X-rays of frequency greater than the *K* limit. If ν_K, ν_M, ν_N are the frequencies corresponding to the *K*, *L*, *M*, ... limits and ν is that of the incident radiation ($\nu > \nu_K$), the velocities v_K, v_M, v_N of the electrons ejected from the *K*, *L*, *M*, ... levels are given by

$$\tfrac{1}{2}mv_K^2 = h(\nu - \nu_K),$$
$$\tfrac{1}{2}mv_L^2 = h(\nu - \nu_L),$$
$$\tfrac{1}{2}mv_M^2 = h(\nu - \nu_M),$$

....

By means of a magnetic field the paths of the electrons are

[1] N. Bohr, 'The Theory of Spectra and Atomic Constitution,' pp. 121–5, 1922.

[2] D. Coster, Phil. Mag. **43**, p. 1070, 1922.

[3] M. de Broglie, Comptes Rendus, **172** and **173** (several papers), 1921; 'Les Rayons X,' Paris, 1922;
Report of the Solvay Congress, 1921, published as 'Atomes et Électrons,' p. 80, Paris, 1923.

bent into arcs of circles of radii proportional to their velocities. Thus 'a corpuscular spectrum' is obtained, the positions of the different images giving the velocities v_K, v_L, v_M, ..., whence ν_K, ν_L, ν_M are found; the spectrum consists of bands terminating sharply on the side of maximum velocity.

107. γ-*Rays*. Ellis[1] has applied this method to the electrons ejected from W, Pt, Pb, U by the γ-rays of radium B. Using the equation $\frac{1}{2}mv_K^2 = h\nu - h\nu_K$, the frequency ν of the γ-rays was determined from the observed value of v_K and the known value of ν_K.

The existence of three γ-rays for which

$$\lambda = 5{\cdot}1,\ 4{\cdot}2,\ 3{\cdot}6 \times 10^{-10} \text{ cm.}$$

was proved; these magnitudes indicate that the γ-rays are due to nuclear disturbances, and the different wave lengths suggest a series of stationary states within the nucleus itself.

[1] C. D. Ellis, Proc. Roy. Soc. **99** A, p. 261, 1921.

On the next two pages is a table of the elements showing Bohr's theory of their atomic structure, which is considered in the chapter which follows.

At. No.	Element	1_1	2_1	2_2	3_1	3_2	3_3	4_1	4_2	4_3	4_4	5_1	5_2
1	H	1	.	.	.	.	.	.	.	.	.	.	.
2	He	2	.	.	.	.	.	.	.	.	.	.	.
3	Li	2	1	.	.	.	.	.	.	.	.	.	.
4	Be	2	2	.	.	.	.	.	.	.	.	.	.
5	B	2	3	.	.	.	.	.	.	.	.	.	.
6	C	2	4	.	.	.	.	.	.	.	.	.	.
7	N	2	4	1	.	.	.	.	.	.	.	.	.
8	O	2	4	2	.	.	.	.	.	.	.	.	.
9	F	2	4	3	.	.	.	.	.	.	.	.	.
10	Ne	2	4	4	.	.	.	.	.	.	.	.	.
11	Na	2	4	4	1	.	.	.	.	.	.	.	.
12	Mg	2	4	4	2	.	.	.	.	.	.	.	.
13	Al	2	4	4	3	.	.	.	.	.	.	.	.
14	Si	2	4	4	4	.	.	.	.	.	.	.	.
15	P	2	4	4	4	1	.	.	.	.	.	.	.
16	S	2	4	4	4	2	.	.	.	.	.	.	.
17	Cl	2	4	4	4	3	.	.	.	.	.	.	.
18	A	2	4	4	4	4	.	.	.	.	.	.	.
19	K	2	4	4	4	4	.	1	.	.	.	.	.
20	Ca	2	4	4	4	4	.	2	.	.	.	.	.
21	Sc	2	4	4		↑		.	.	.	.	.	.
22	Ti	2	4	4				.	.	.	.	.	.
23	V	2	4	4				.	.	.	.	.	.
24	Cr	2	4	4				.	.	.	.	.	.
25	Mn	2	4	4				.	.	.	.	.	.
26	Fe	2	4	4				.	.	.	.	.	.
27	Co	2	4	4		↓		.	.	.	.	.	.
28	Ni	2	4	4	6	6	6	.	.	.	.	.	.
29	Cu	2	4	4	6	6	6	1	.	.	.	.	.
30	Zn	2	4	4	6	6	6	2	.	.	.	.	.
31	Ga	2	4	4	6	6	6	3	.	.	.	.	.
32	Ge	2	4	4	6	6	6	4	.	.	.	.	.
33	As	2	4	4	6	6	6	4	1	.	.	.	.
34	Se	2	4	4	6	6	6	4	2	.	.	.	.
35	Br	2	4	4	6	6	6	4	3	.	.	.	.
36	Kr	2	4	4	6	6	6	4	4	.	.	.	.
37	Rb	2	4	4	6	6	6	4	4	.	.	1	.
38	Sr	2	4	4	6	6	6	4	4	.	.	2	.
39	Y	2	4	4	6	6	6			↑		.	.
40	Zr	2	4	4	6	6	6					.	.
41	Nb	2	4	4	6	6	6					.	.
42	Mo	2	4	4	6	6	6					.	.
43	—	2	4	4	6	6	6					.	.
44	Ru	2	4	4	6	6	6					.	.
45	Rh	2	4	4	6	6	6			↓		.	.
46	Pd	2	4	4	6	6	6	6	6	6	.	.	.
47	Ag	2	4	4	6	6	6	6	6	6	.	1	.
48	Cd	2	4	4	6	6	6	6	6	6	.	2	.
49	In	2	4	4	6	6	6	6	6	6	.	3	.
50	Sn	2	4	4	6	6	6	6	6	6	.	4	.
51	Sb	2	4	4	6	6	6	6	6	6	.	4	1
52	Te	2	4	4	6	6	6	6	6	6	.	4	2
53	I	2	4	4	6	6	6	6	6	6	.	4	3
54	Xe	2	4	4	6	6	6	6	6	6	.	4	4

At. No.	Element	1_1	2_1	2_2	3_1	3_2	3_3	4_1	4_2	4_3	4_4	5_1	5_2	5_3	5_4	6_1	6_2	6_3	7_1	7_2
55	Cs	2	4	4	6	6	6	6	6	6	.	4	4	.	.	1	.	.	.	.
56	Ba	2	4	4	6	6	6	6	6	6	.	4	4	.	.	2	.	.	.	.
57	La	2	4	4	6	6	6									.	.	.	.	.
58	Ce	2	4	4	6	6	6									.	.	.	.	.
59	Pr	2	4	4	6	6	6									.	.	.	.	.
60	Nd	2	4	4	6	6	6									.	.	.	.	.
61	—	2	4	4	6	6	6									.	.	.	.	.
62	Sm	2	4	4	6	6	6									.	.	.	.	.
63	Eu	2	4	4	6	6	6									.	.	.	.	.
64	Gd	2	4	4	6	6	6									.	.	.	.	.
65	Tb	2	4	4	6	6	6									.	.	.	.	.
66	Ds	2	4	4	6	6	6									.	.	.	.	.
67	Ho	2	4	4	6	6	6									.	.	.	.	.
68	Er	2	4	4	6	6	6									.	.	.	.	.
69	Tm	2	4	4	6	6	6									.	.	.	.	.
70	Yb	2	4	4	6	6	6									.	.	.	.	.
71	Lu	2	4	4	6	6	6									.	.	.	.	.
72	—	2	4	4	6	6	6									.	.	.	.	.
73	Ta	2	4	4	6	6	6									.	.	.	.	.
74	W	2	4	4	6	6	6									.	.	.	.	.
75	—	2	4	4	6	6	6									.	.	.	.	.
76	Os	2	4	4	6	6	6									.	.	.	.	.
77	Ir	2	4	4	6	6	6									.	.	.	.	.
78	Pt	2	4	4	6	6	6	8	8	8	8	6	6	6	.	.	.	.	.	.
79	Au	2	4	4	6	6	6	8	8	8	8	6	6	6	.	1	.	.	.	.
80	Hg	2	4	4	6	6	6	8	8	8	8	6	6	6	.	2	.	.	.	.
81	Tl	2	4	4	6	6	6	8	8	8	8	6	6	6	.	3	.	.	.	.
82	Pb	2	4	4	6	6	6	8	8	8	8	6	6	6	.	4	.	.	.	.
83	Bi	2	4	4	6	6	6	8	8	8	8	6	6	6	.	4	1	.	.	.
84	Po	2	4	4	6	6	6	8	8	8	8	6	6	6	.	4	2	.	.	.
85	—	2	4	4	6	6	6	8	8	8	8	6	6	6	.	4	3	.	.	.
86	Nt	2	4	4	6	6	6	8	8	8	8	6	6	6	.	4	4	.	.	.
87	—	2	4	4	6	6	6	8	8	8	8	6	6	6	.	4	4	.	1	.
88	Ra	2	4	4	6	6	6	8	8	8	8	6	6	6	.	4	4	.	2	.
89	Ac	2	4	4	6	6	6	8	8	8	8								.	.
90	Th	2	4	4	6	6	6	8	8	8	8								.	.
91	Pa	2	4	4	6	6	6	8	8	8	8								.	.
92	U	2	4	4	6	6	6	8	8	8	8								.	.
110	—	2	4	4	6	6	6	8	8	8	8	8	8	8	8	6	6	6	.	.
111	—	2	4	4	6	6	6	8	8	8	8	8	8	8	8	6	6	6	1	.
112	—	.	.	.	.	.	.	.	.	.	.	.	.	.	.	.	.	.	2	.
113	—	.	.	.	.	.	.	.	.	.	.	.	.	.	.	.	.	.	3	.
114	—	.	.	.	.	.	.	.	.	.	.	.	.	.	.	.	.	.	4	.
115	—	.	.	.	.	.	.	.	.	.	.	.	.	.	.	.	.	.	4	1
116	—	.	.	.	.	.	.	.	.	.	.	.	.	.	.	.	.	.	4	2
117	—	.	.	.	.	.	.	.	.	.	.	.	.	.	.	.	.	.	4	3
118	—	2	4	4	6	6	6	8	8	8	8	8	8	8	8	6	6	6	4	4

CHAPTER XVI

BOHR'S THEORY OF THE ATOMIC STRUCTURE OF THE ELEMENTS

108. *Bohr's theory of the atomic structure of the elements.* The elements are considered in the order of their atomic number beginning with hydrogen, and in the groups into which they fall in the periodic table of Mendeléeff.

109. *First period. Hydrogen* (1)—*Helium* (2) [the bracket after the symbol of an element will contain its atomic number]. The work of Bohr[1], Landé, and Kramers indicates a 1_1 orbit for the electron of H and 1_1 orbits for the electrons of He. This completes the *K* shell.

110. *Second period. Lithium* (3)—*Neon* (10). The absorption spectrum of Li (3) indicates a 2_1 orbit for the third electron. As regards Be (4), B (5), C (6) there are no simple series spectra known. Bohr assumes on the ground of general physical and chemical evidence that the fourth, fifth[2] and

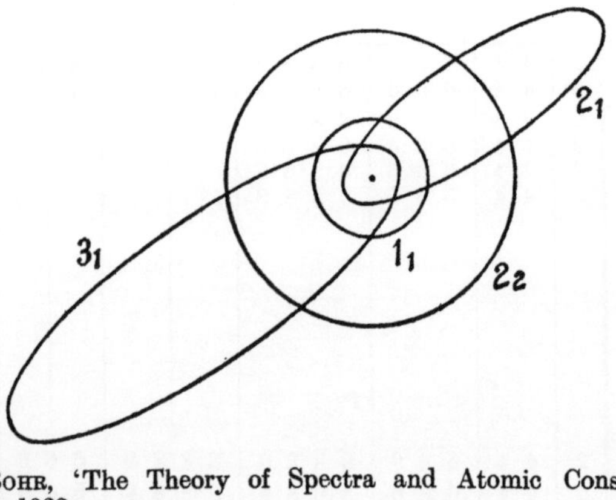

[1] N. Bohr, 'The Theory of Spectra and Atomic Constitution,' pp. 86–89, 1922.

[2] There is however evidence from the spectrum of C_+, that the orbit of the 5th electron in B is a 2_2 one.

sixth electrons are all bound in 2_1 orbits, so that in the C atom the four 2_1 orbits are probably arranged with the normals to their planes symmetrically spaced in direction like the lines from the centre to the vertices of a regular tetrahedron. These 2_1 orbits penetrate at perihelion into the region of 1_1 orbits. The seventh electron in N (7) is not likely to be in a 2_1 orbit, as that would destroy the symmetry of the interaction of these electrons. Bohr assumes that the new electrons of N (7), O (8), F (9), Ne (10) are all in 2_2 orbits. In neon, the planes of the four 2_2 orbits must have a high degree of spatial symmetry in harmony with that of the four 2_1 orbits. The above would allow for the great difference in the properties of the elements in the two halves of the second period. The L shell is completed with neon, an inert gas.

111. *Third period. Sodium* (11)—*Argon* (18). The analogy of Na to Li suggests the beginning of a new shell with a 3_1 orbit for the 11th electron of sodium[1].

For Mg (12) the arc spectrum shows that the ground orbit is a 3_1 orbit. For Al (13) the absorption spectrum corresponds not to the principal series (as is usually the case) (§ 102) but to the sharp and diffuse series. Hence the term $2p$ is the normal state, not the term $1s$, so that the 13th electron is in a 3_2 orbit (cf. the possible 2_2 orbit for the 5th electron in B). But for Si (14) we assume four 3_1 orbits symmetrically arranged as in C. Then as in the second period, the next four electrons are in 3_2 orbits symmetrically arranged, so that the eight 3 orbits are grouped in A (18) just as the eight 2 orbits in Ne (10). This picture shows up the similarity in the chemical properties of the elements in the corresponding parts of the second and third periods. The M shell is completed with the formation of argon, again an inert gas.

112. *Fourth period. Potassium* (19)—*Krypton* (36). K (19), Ca (20) correspond chemically to Na (11), Mg (12); the spectrum of K indicates a 4_1 orbit.

[1] N. Bohr, loc. cit. pp. 95–98.

But the spectrum of K (associated with the binding of the 19th electron) shows that the energy of the 4_1 state (1*s*) is much greater than that of the 3_3 state (3*d*); while the spectrum of Ca_+ (associated with the binding of the 19th electron in Ca) shows that the energy of the 4_1 state (1*s*) is only a little greater than that of the 3_3 state (3*d*) (see Bohr's figure below). Hence for the next element Sc (21) we must be prepared to find that a 3_3 orbit will have greater energy (in all this the negative energy W is meant) than a 4_1 orbit, that is, it will correspond to a stronger binding to the atom. For the next elements Sc (21)—Ni (28) there is a further development of the 3 orbits (an *inner* group) (see the table on p. 156), and only when this

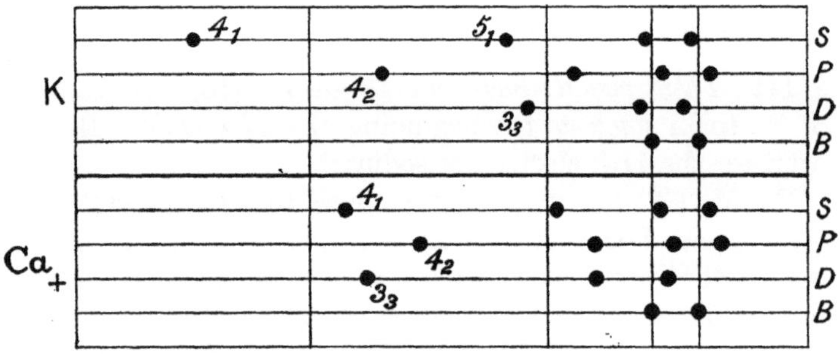

is complete at Ni can we expect to find once more the change of properties of the elements with increasing atomic number due to the growth of the *outer* shell.

The whole group of 18 elements is complete at Kr (36), which we expect to have four 4_1 orbits and four 4_2 orbits symmetrically arranged as its outer shell, as in A and Ne. This would require the 3-quantum groups to contain 18 electrons, and this suggests three groups of six electrons in 3_1, 3_2, 3_3 orbits. It is possible to arrange three configurations each having six orbits in a simple manner, each group having trigonal symmetry.

The spectrum of Cu (29) suggests that in it the 3-quantum orbits are complete and a single 4_1 orbit is outside (table on p. 156), for in contrast to the very complicated spectra of the preceding elements (which include iron, cobalt, nickel, with notorious spectra) resulting from the unsettled character of the 3-quantum orbits, it has a simple spectrum very much like that of sodium.

The elements Sc (21)—Ni (28) correspond to changes in the 3-quantum orbits from a symmetrical configuration of eight in two groups to eighteen in three groups; this corresponds to the development of the magnetic properties, which completely disappear with the completion of the group of eighteen at Cu. Bohr suggests that the investigation of the magnetic properties of the elements may give a clue as to the way in which the change from 8 to 18 in the 3-quantum orbits occurs. Also the striking colours shown by the salts of these metals (Sc—Ni) are connected with this view; colour, that is, the absorption of light in the visible part of the spectrum, is evidence of transitions involving energy changes of the same order as in optical spectrum theory—these transitions even under normal conditions may occur between 3_3 and 4_1 orbits when the balance of energy between them is small (cf. the figure of the K, Ca_+ spectra, p. 160) before the 6, 6, 6 arrangement is stabilised.

113. *Fifth period. Rubidium* (37)—*Xenon* (54). For Rb (37), Sr (38) we assume 5_1 orbits for the new electrons. This is supported by the spectra of Rb and Sr_+. These however indicate (as for K and Ca_+) that 4_3 orbits will soon appear, and from Y (39) to Pd (46) there is a stage where the 4-quantum orbits change from 4, 4 to 6, 6, 6 (see the table on p. 156). At the end of this period Xe (54) has in its outer shell the symmetrical arrangement of eight 5-quantum orbits like the outer ones of the other inert gases Kr, A, Ne.

114. *Sixth period. Caesium* (55)—*Niton* (86). Cs (55), Ba (56) have new electrons in 6_1 orbits; the spectra confirm this. After that a 5_3 orbit becomes a firmer binding than a 6_1 orbit, and also a stage is reached (as successive electrons are bound) when 4_4 orbits are firmer than either a 5_3 or a 6_1 orbit. Hence from La (57)—Pt (78) there is a rearrangement both of the 4- and 5-quantum groups, which at the beginning consist of 6, 6, 6, 4-quantum orbits and 4, 4, 5-quantum orbits and at the end consist of 8, 8, 8, 8, 4-quantum orbits and 6, 6, 6, 5-quantum orbits. These are the *rare earths*. Again there are similar chemical properties in these rare earths, as the *outer* electrons are not being added to, and there are magnetic and colour changes of an extreme kind just as in the group Sc—Ni in the fourth period, where changes in the inner group of 3-quantum orbits were taking place. The radio-active chemically-inactive gas niton has the outer shell of four 6_1 and four 6_2 orbits arranged with the same space configuration as the other inert gases.

115. *Seventh period.* (87)—(118). This is for the great part an ideal continuation of the building up of atoms up to this point. Only the elements Ra (88)—U (92) are known; 87 has not been discovered and U is the heaviest element known. In the neutral atom of Ra there would be the structure of Nt and two electrons in 7_1 orbits. Then the 6_3 orbits would compete with the 7_1 ones and the 5_4 ones with both of these, so that we should expect the elements from Ac (89)—(110) to have the 5- and 6-quantum orbits changing from sets of 6, 6, 6 and 4, 4 to 8, 8, 8, 8 and 6, 6, 6 (see the table on p. 157). After that there would be the outer shell of four 7_1

Element	At. No.	No. of electrons in n_k orbits																						
		1_1	2_1	2_2	3_1	3_2	3_3	4_1	4_2	4_3	4_4	5_1	5_2	5_3	5_4	5_5	6_1	6_2	6_3	6_4	6_5	6_6	7_1	7_2
Helium	2	2	.	.	.	.	.	.	.	.	.	.	.	.	.	.	.	.	.	.	.	.	.	.
Neon	10	2	4	4	.	.	.	.	.	.	.	.	.	.	.	.	.	.	.	.	.	.	.	.
Argon	18	2	4	4	4	4	.	.	.	.	.	.	.	.	.	.	.	.	.	.	.	.	.	.
Krypton	36	2	4	4	6	6	6	4	4	.	.	.	.	.	.	.	.	.	.	.	.	.	.	.
Xenon	54	2	4	4	6	6	6	6	6	6	.	4	4	.	.	.	.	.	.	.	.	.	.	.
Niton	86	2	4	4	6	6	6	8	8	8	8	6	6	6	.	.	4	4	.	.	.	.	.	.
?	118	2	4	4	6	6	6	8	8	8	8	8	8	8	8	.	6	6	6	.	.	.	4	4

and four 7_2 orbits for the inert gas (118), as for Nt, Xe, Kr, A, Ne.

The inert gases exhibit the stable grouping of the orbits when the outermost group is just complete.

The table on p. 162 shows at a glance the nature of the changes of the inner groups, as we pass through period after period of the table.

CHAPTER XVII

BAND SPECTRA IN THE INFRA-RED

116. *Band spectra.* These are spectra in which there is a very close sequence of lines; their origin is the rotation and vibration of the *molecule.* A molecule is a system of atoms each having a nucleus with its attendant electrons.

From the theory of the specific heats of gases[1] it is known that at ordinary temperatures the effect of the rotation of the molecules comes fully into play; the vibrations of the nuclei of the atoms relative to one another, at any rate for the simpler gases, oxygen, nitrogen and the like, are not appreciable at ordinary temperatures, but show themselves in their effect on the specific heat when the gas is strongly heated; disturbances of the electron orbits relative to the nuclei also appear at very high temperatures.

In the following theory the effect of rotation will first be considered by treating the molecule as a rigid body. Then the effect of internal vibration will be taken into account by treating the atoms as point-masses oscillating relative to each other as they rotate, and afterwards allowing for the fact that the atoms are not points, but nuclei surrounded by electrons.

117. *The infra-red absorption bands.*

(i) *Rotation spectrum.* Consider a diatomic molecule, such as that of HCl (with two atoms H_+, Cl_-). Suppose it to consist of two point-masses m_1, m_2 at a constant distance

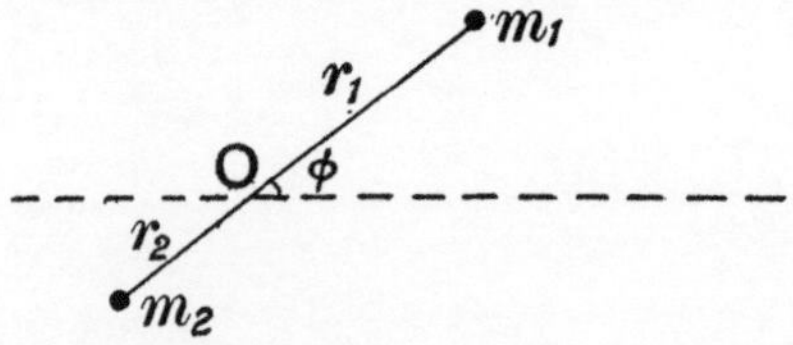

[1] G. Birtwistle, 'The Principles of Thermodynamics,' Cambridge, p. 135, 1925.

a apart with charges e, $-e$, the system revolving in one plane with uniform angular velocity $\dot{\phi}$ about the centre of gravity O.

The kinetic energy

$$T = \tfrac{1}{2}m_1 r_1^2\dot{\phi}^2 + \tfrac{1}{2}m_2 r_2^2\dot{\phi}^2$$
$$= \tfrac{1}{2}Ma^2\dot{\phi}^2,$$

where $$a = r_1 + r_2,$$

and $$\frac{1}{M} = \frac{1}{m_1} + \frac{1}{m_2}.$$

The momentum $$p = \frac{\partial T}{\partial \dot{\phi}} = Ma^2\dot{\phi}.$$

Therefore the Hamiltonian function

$$H = \frac{p^2}{2Ma^2} = \alpha_1, \text{ say.}$$

Now $$I = \oint p\,d\phi = \int_0^{2\pi} \sqrt{2Ma^2\alpha_1}\,d\phi = 2\pi\sqrt{2Ma^2\alpha_1},$$

$$\therefore\ \alpha_1 = \frac{I^2}{8\pi^2 Ma^2} = \frac{I^2}{8\pi^2 A},$$

where A is the moment of inertia of the system about a line through O perpendicular to the line joining m_1 and m_2.

The frequency of revolution

$$\omega = \frac{\partial H}{\partial I} = \frac{\partial \alpha_1}{\partial I} = \frac{I}{4\pi^2 A}.$$

The quantum condition is $I = mh$, where m is an integer, so that $\alpha_1 = \dfrac{m^2h^2}{8\pi^2 A}$; and α_1 is the total energy.

For a transition $m_1 \rightarrow m_2$,

$$h\nu = \alpha_1 - \alpha_2 = \frac{h^2}{8\pi^2 A}(m_1^2 - m_2^2).$$

Now the electric moment of the molecule has components

$$M_x = \Sigma ex = ea\cos\phi,$$
$$M_y = \Sigma ey = \pm\, ea\sin\phi,$$
$$\therefore\ M_x + iM_y = ea\epsilon^{\pm i\phi} = ea\epsilon^{\pm i(2\pi\omega t)},$$

where ϵ is the base of logarithms. Thus by the correspondence principle (since the index is $2\pi i\tau\omega t$, where $\tau = \pm 1$), the quantum number m can only change by one unit.

For a transition $m + 1 \rightarrow m$ (which corresponds to an *emission* of a line),

$$\nu = \frac{h}{8\pi^2 A}\{(m+1)^2 - m^2\} = \frac{h^2}{8\pi^2 A}(2m+1).$$

Thus the spectrum consists of a series of lines at equal distances $\dfrac{h}{4\pi^2 A}$.

Corresponding to $m = 1$, $\nu = \dfrac{3h}{8\pi^2 \mu a^2}$.

a is known to be of order 1 Å.; let it $= k$ Å.

M is approximately, for HCl, equal to m_H, since $m_{Cl} = 35 m_H$.

$$\therefore\ \nu = \frac{3 \times 6{\cdot}55 \times 10^{-27}}{8\pi^2 \times 1{\cdot}66 \times 10^{-24} \times (k \times 10^{-8})^2} \quad \text{and} \quad \lambda = \frac{c}{\nu}.$$

Therefore

λ is about $2 \times 10^{-2} k^2$ cm. or $(200k^2)\ \mu$, where $\mu = 10^{-4}$ cm.

This line is far in the infra-red. Pure rotation spectra have been observed in the case of water vapour $\left(\begin{matrix} H_+ \\ H_+ \end{matrix} O_{--}\right)$ by Rubens; they lie in the region $100\,\mu$.

118. (ii) *Rotation-vibration spectrum.* We now consider the effect of the vibrations of the atoms, neglecting electron disturbances. Regard the diatomic molecule as dynamically equivalent to two point masses with an elastic connection which permits the nuclei (the effective masses of the atoms) to oscillate along the line joining them. Let a be the distance apart of the nuclei if there is no rotation, r_0 their distance when rotating uniformly with angular momentum α_2. For their motion in general

$$\left.\begin{aligned} T &= \tfrac{1}{2} M (\dot{r}^2 + r^2\dot{\theta}^2) \\ V &= f(r) \end{aligned}\right\},$$

where

$$\frac{1}{M} = \frac{1}{m_1} + \frac{1}{m_2}.$$

[As there is stable equilibrium with $r = a$, $\dot{\theta} = 0$, then $f'(a) = 0$, and $f''(a)$ is positive.]

Now $$p_1 = \frac{\partial T}{\partial \dot{r}} = M\dot{r}, \quad p_2 = \frac{\partial T}{\partial \dot{\theta}} = Mr^2\dot{\theta},$$

so that $$H = T + V = \tfrac{1}{2}M(\dot{r}^2 + r^2\dot{\theta}^2) + f(r)$$
$$= \frac{1}{2M}\left(p_1^2 + \frac{p_2^2}{r^2}\right) + f(r).$$

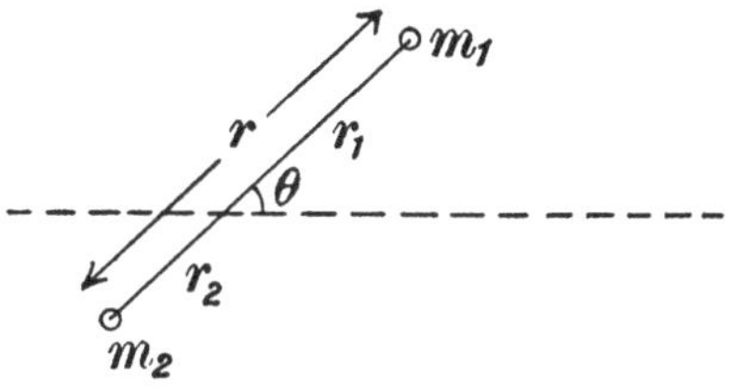

Also $\frac{\partial H}{\partial \theta} = \dot{p}_2$. $\therefore$ $\dot{p}_2 = 0$ and $p_2 =$ constant, and, being the angular momentum, is equal to the α_2 mentioned above.

$$\therefore \; p_2 = \alpha_2.$$

Again $$\frac{\partial H}{\partial r} = \dot{p}_1,$$

$$\therefore \; -\frac{p_2^2}{Mr^3} + f'(r) = M\ddot{r},$$

or $$M\ddot{r} = -\frac{\alpha_2^2}{Mr^3} + f'(r).$$

Since $\ddot{r} = 0$, when $r = r_0$,

$$Mr_0^3 f'(r_0) = \alpha_2^2 \quad \text{.....................(1).}$$

Let $r_0 = a + \rho$, where ρ is small (the force between the atoms being supposed large); then, neglecting ρ^2, ...

$$M(a + \rho)^3 f'(a + \rho) = \alpha_2^2,$$

$$\therefore \; Ma^3\rho f''(a) = \alpha_2^2, \text{ since } f'(a) = 0,$$

$$\therefore \; \rho = \alpha_2^2 / Ma^3 f''(a) \quad \text{..................(2).}$$

Now $$H = \frac{1}{2M}\left(p_1^2 + \frac{p_2^2}{r^2}\right) + f(r).$$

Writing $$r = r_0 + x,$$
so that $$r = a + \rho + x,$$
we have
$$H = \frac{1}{2M}\left\{p_1^2 + \frac{\alpha_2^2}{(a+\rho+x)^2}\right\} + f(a+\rho+x).$$

Neglecting squares of ρ and terms such as $\rho\alpha_2^2$, on account of (2),
$$H = \frac{1}{2M}\left\{p_1^2 + \frac{\alpha_2^2}{(a+x)^2}\right\} + f(a+\rho+x), \text{ to this order,}$$
$$= \frac{1}{2M}\left\{p_1^2 + \frac{\alpha_2^2}{a^2}\left(1 - \frac{2x}{a} + \frac{3x^2}{a^2} - \frac{4x^3}{a^3}\right)\right\} + f(a) + (\rho + x) f'(a)$$
$$+ \frac{(x^2 + 2\rho x)}{2} f''(a) + \frac{x^3 + 3\rho x^2}{6} f'''(a) + \ldots,$$
and $$f'(a) = 0,$$
$$\therefore\ H = \frac{p_1^2}{2M} + \left(f + \frac{\alpha_2^2}{2Ma^2}\right) + x^2\left(\tfrac{1}{2}f'' + \frac{3\alpha_2^2}{2Ma^4} + \rho\frac{f'''}{2}\right)$$
$$+ x^3\left(\frac{f'''}{6} - \frac{2\alpha_2^2}{Ma^5}\right) + \ldots,$$
the coefficient of x being zero, on account of (2), or
$$H = \frac{p_1^2}{2M} + \left(f + \frac{\alpha_2^2}{2Ma^2}\right) + x^2\left\{\tfrac{1}{2}f'' + \frac{\alpha_2^2}{2M}\left(\frac{3}{a^4} + \frac{f'''}{a^3 f''}\right)\right\}$$
$$+ x^3\left\{\frac{f'''}{6} - \frac{2\alpha_2^2}{Ma^5}\right\}, \text{ using (2).}$$
$$\therefore\ H = c_0 + \frac{p_1^2}{2M} + \tfrac{1}{2}M(2\pi\omega)^2 x^2 + cx^3 \quad \ldots\ldots(3),$$

where
$$\left.\begin{aligned} c_0 &= f + \frac{\alpha_2^2}{2Ma^2} \\ \tfrac{1}{2}M(2\pi\omega)^2 &= \tfrac{1}{2}f'' + \frac{\alpha_2^2}{2M}\left(\frac{3}{a^4} + \frac{f'''}{a^3 f''}\right) \\ c &= \tfrac{1}{6}f''' - \frac{2\alpha_2^2}{Ma^5} \end{aligned}\right\}, \text{ define } c_0, \omega, c.$$

Also $$p_1 = \frac{\partial T}{\partial \dot{r}} = \frac{\partial T}{\partial \dot{x}}, \quad \text{since } r = a + \rho + x.$$

If c were zero, (3) would lead to a simple harmonic motion of frequency ω;

with c not zero we have the so-called *anharmonic oscillator*.

Writing $H = \alpha_1$, we have

$$c_0 + \frac{p_1^2}{2M} + \tfrac{1}{2}M(2\pi\omega)^2 x^2 + cx^3 + \ldots = \alpha_1,$$

$$\therefore \; p_1^2 = 2M(\alpha_1 - c_0) - (2\pi M\omega)^2 x^2 - 2Mcx^3, \text{ to this order.}$$

$$\therefore \; I_1 = \oint p_1 dx = \oint dx \sqrt{2M(\alpha_1 - c_0) - (2\pi M\omega)^2 x^2 - 2Mcx^3}$$

$$= 2\pi M\omega \oint dx \sqrt{a^2 - x^2 - \lambda x^3},$$

where $$a^2 = \frac{\alpha_1 - c_0}{2M(\pi\omega)^2}, \quad \lambda = \frac{c}{2M(\pi\omega)^2}.$$

Also $$I_2 = \oint p_2 d\theta = \int_0^{2\pi} \alpha_2 d\theta = 2\pi\alpha_2.$$

119. *Calculation of* $\oint \sqrt{a^2 - x^2 - \lambda x^3}\, dx$, *neglecting* λ^3.... To a first approximation the roots of $a^2 - x^2 - \lambda x^3 = 0$ are those of $a^2 - x^2 = 0$ or are $x = \pm a$. Writing $x = a + \xi$ in the equation, we have

$$-2\xi a - \xi^2 - \lambda(a^3 + 3a^2\xi + 3a\xi^2 + \xi^3) = 0,$$

and the most important terms give $\xi = -\lambda a^2/2$. Using this in the other terms we have

$$-2\xi a = \frac{\lambda^2 a^4}{4} + \lambda a^3 - 3a^2\lambda\frac{\lambda a^2}{2}, \text{ approx.,}$$

$$\therefore \; \xi = \frac{5}{8}\lambda^2 a^3 - \frac{\lambda a^2}{2}.$$

$\therefore$ if x_1, x_2, x_3 are the roots of the cubic,

$$x_1 = a - \frac{\lambda a^2}{2} + \frac{5}{8}\lambda^2 a^3, \text{ approx.}$$

So $$x_2 = -a - \frac{\lambda a^2}{2} - \frac{5}{8}\lambda^2 a^3.$$

Since
$$x_1 + x_2 + x_3 = -\frac{1}{\lambda},$$
we have
$$x_3 = -\frac{1}{\lambda} + \lambda a^2 \ldots .$$
Now

$$\oint \sqrt{a^2 - x^2 - \lambda x^3}\, dx = \oint \sqrt{\lambda} \sqrt{(x_1 - x)(x - x_2)(x - x_3)}\, dx$$
$$= \oint \sqrt{-\lambda x_3} \{\sqrt{(x_1 - x)(x - x_2)}\} \left\{1 - \frac{x}{2x_3} - \frac{x^2}{8{x_3}^2} \ldots\right\} dx.$$

Writing $x = x_1 \sin^2 \phi + x_2 \cos^2 \phi$,

$$\oint \sqrt{(x_1 - x)(x - x_2)}\, dx$$
$$= 2\int_{x_2}^{x_1} \sqrt{(x_1 - x)(x - x_2)}\, dx$$
$$= 2\int_0^{\frac{\pi}{2}} 2(x_1 - x_2)^2 \sin^2\phi \cos^2\phi\, d\phi$$
$$= \frac{\pi}{4}(x_1 - x_2)^2 \equiv U, \text{ suppose.}$$

So
$$\oint x\sqrt{(x_1 - x)(x - x_2)}\, dx = \tfrac{1}{2}(x_1 + x_2)\, U,$$
and
$$\oint x^2 \sqrt{(x_1 - x)(x - x_2)}\, dx = \frac{1}{16}[5(x_1 + x_2)^2 - 4x_1x_2]\, U.$$

$$\therefore \oint \sqrt{a^2 - x^2 - \lambda x^3}\, dx$$
$$= \sqrt{-\lambda x_3}\, U \left[1 - \frac{x_1 + x_2}{4x_3} - \frac{\{5(x_1 + x_2)^2 - 4x_1x_2\}}{128{x_3}^2}\right]$$
$$= \sqrt{1 - \lambda^2 a^2}\, U \left[1 - \frac{\lambda^2 a^2}{4} - \frac{\lambda^2}{128}(4a^2)\right],$$

inserting the values of x_1, x_2, x_3 and neglecting λ^3, λ^4, ...;

$$= \left(1 - \frac{\lambda^2 a^2}{2}\right)\frac{\pi}{4}\left(2a + \frac{5}{4}\lambda^2 a^3\right)^2 \left[1 - \frac{9\lambda^2 a^2}{32}\right]$$
$$= \pi a^2 \left(1 - \frac{\lambda^2 a^2}{2}\right)\left(1 + \frac{5}{4}\lambda^2 a^2\right)\left(1 - \frac{9}{32}\lambda^2 a^2\right)$$
$$= \pi a^2 \left(1 + \frac{15}{32}\lambda^2 a^2\right).$$

120. Hence

$$\left.\begin{aligned} I_1 &= 2\pi M\omega\pi a^2\left(1+\frac{15}{32}\lambda^2 a^2\right) \\ I_2 &= 2\pi\alpha_2 \end{aligned}\right\}.$$

We require a^2, which contains α_1.

Neglecting λ, $a^2 = I_1/2\pi^2 M\omega$, and using this approximation in the term containing λ,

$$I_1 = 2\pi^2 M\omega a^2 + \frac{15}{32}\lambda^2 \frac{I_1^2}{2\pi^2 M\omega},$$

$$\therefore\ a^2 = \frac{I_1}{2\pi^2 M\omega} - \frac{15}{32}\lambda^2 \frac{I_1^2}{(2\pi^2 M\omega)^2} \quad \ldots\ldots\ldots(4).$$

Now $$a^2 = \frac{\alpha_1 - c_0}{2M(\pi\omega)^2},$$

$$\therefore\ \alpha_1 = c_0 + 2M(\pi\omega)^2 a^2$$

$$= c_0 + \omega I_1 - \frac{15}{32}\lambda^2 \frac{I_1^2}{2\pi^2 M}.$$

Therefore, inserting the value of λ^2, we find

$$\alpha_1 = c_0 + \omega I_1 - \frac{15c^2 I_1^2}{16\pi^2(2\pi\omega)^4 M^3} \quad \ldots\ldots\ldots\ldots(5).$$

Now $$\omega^2 = \frac{1}{2\pi^2 M}\left\{\tfrac{1}{2}f'' + \frac{\alpha_2^2}{2M}\left(\frac{3}{a^4} + \frac{f'''}{a^3 f''}\right)\right\},$$

$$\therefore\ \omega = \frac{1}{2\pi}\sqrt{\frac{f''}{M}}\,(1 + k\alpha_2^2),$$

where k is a constant.

If ν_0 is the frequency of the small oscillations if the molecule were at rest, ν_0 is found from the equation

$$M\ddot{r} = -f'(r),$$

where $$r = a + x,$$

so that $$M\ddot{x} = -f'(a+x) = -xf''(a),$$

since $$f'(a) = 0.$$

$$\therefore\ \nu_0 = \frac{1}{2\pi}\sqrt{\frac{f''(a)}{M}},$$

$$\therefore\ \omega = \nu_0(1 + k\alpha_2^2),$$

and using $\alpha_2 = I_2/2\pi$, we obtain $\omega = \nu_0 + \beta I_2^2$, where β is a constant for the molecule.

In the last term of (5), which is a small correction, we may neglect α_2^2, so that c^2/ω^4 is a constant for the molecule, being equal to $f'''^2/36\nu_0^2$ (from the equations at the foot of p. 168). Therefore the term is γI_1^2, where γ is constant for the molecule.

Also $$c_0 = f(a) + \alpha_2^2/2Ma^2,$$

$$\therefore \ \alpha_1 = f(a) + \frac{I_2^2}{8\pi^2 Ma^2} + (\nu_0 + \beta I_2^2) I_1 + \gamma I_1^2, \text{ from (5)},$$

or $$\alpha_1 = f(a) + \frac{I_2^2}{8\pi^2 A_0} + (\nu_0 + \beta I_2^2) I_1 + \gamma I_1^2,$$

where A_0 is the moment of inertia for the molecule at rest.

The quantum conditions are $I_1 = nh$, $I_2 = mh$; so that the energy E $(= \alpha_1)$ of the stationary states is given by

$$E = \epsilon + \frac{m^2h^2}{8\pi^2 A_0} + nh(\nu_0 + \beta m^2h^2) + \gamma n^2h^2 \text{ ......(6)},$$

where ϵ is a constant.

This result is due to Kratzer[1].

If ν is the frequency corresponding to a transition

$$n_1 \rightarrow n_2, \quad m \pm 1 \rightarrow m, \quad \text{then}$$

$$h\nu = \frac{h^2}{8\pi^2 A_0} \{(m \pm 1)^2 - m^2\} + h\nu_0 (n_1 - n_2) + \beta h^3 [n_1 (m \pm 1)^2 - n_2 m^2] + \gamma h^2 [n_1^2 - n_2^2],$$

or

$$\nu = \nu_0 (n_1 - n_2) + \frac{h}{8\pi^2 A_0} (\pm 2m + 1) + \beta h^2 [(n_1 - n_2) m^2 \pm 2mn_1 + n_1] + \gamma h [n_1^2 - n_2^2] \text{ ...(7)},$$

or $$\nu = a \pm bm + cm^2 \text{ ....(8)},$$

where

$$\left.\begin{aligned} a &= \frac{h}{8\pi^2 A_0} + \nu_0 (n_1 - n_2) + \beta h^2 n_1 + \gamma h (n_1^2 - n_2^2) \\ b &= \frac{h}{4\pi^2 A_0} + \beta h^2 n_1 \\ c &= \beta h^2 (n_1 - n_2) \end{aligned}\right\}.$$

[1] A. Kratzer, Zeitschr. für Physik, 3, p. 289, 1920.

121. *The angle variables.*

$$S = \int \frac{\partial S}{\partial r} dr + \int \frac{\partial S}{\partial \theta} d\theta$$

$$= \int p_1 dr + \int p_2 d\theta$$

$$= \alpha_2 \theta + 2\pi M \omega \int dx \sqrt{a^2 - x^2 - \lambda x^3}.$$

Neglecting $\alpha_2{}^2$ to a first approximation, we have

$$a^2 = \frac{\alpha_1 - f(a)}{2M(\pi\omega)^2}, \quad \lambda = \frac{f'''}{6(2M)(\pi\omega)^2}.$$

Also $\alpha_1 = f(a) + \nu_0 I_1 + \gamma I_1{}^2$

$= f(a) + \nu_0 I_1$, approx., since γ is small compared with ν_0.

Also $\omega = \nu_0$ to this order.

$$\therefore \quad a^2 = \nu_0 I_1 / 2M(\pi\nu_0)^2, \quad \lambda = \frac{f'''}{12M\pi^2\nu_0{}^2} \quad \text{.........(9).}$$

To this order, the second term in S is independent of I_2, which occurs only in the first term $\alpha_2\theta$ which is equal to $I_2\theta/2\pi$.

$$\therefore \quad w_2 = \frac{\partial S}{\partial I_2} = \frac{\theta}{2\pi}$$

to this order; or $\theta = 2\pi w_2$,

and
$$w_1 = \frac{\partial S}{\partial I_1} = 2\pi M \nu_0 \int \frac{dx}{2\sqrt{a^2 - x^2 - \lambda x^3}} \frac{\partial}{\partial I_1}(a^2)$$

$$= 2\pi M \nu_0 \frac{\nu_0}{2M(\pi\nu_0)^2} \int \frac{dx}{2\sqrt{a^2 - x^2 - \lambda x^3}}, \text{ using (9)}$$

$$\therefore \quad 2\pi w_1 = \int \frac{dx}{\sqrt{a^2 - x^2 - \lambda x^3}}$$

$$= \int \frac{dx}{\sqrt{\lambda(x_1 - x)(x - x_2)(x - x_3)}}$$

$$= \frac{1}{\sqrt{-\lambda x_3}} \int \frac{dx}{\sqrt{(x_1 - x)(x - x_2)}} \left(1 + \frac{x}{2x_3} + \ldots\right).$$

Writing $x = x_1 \sin^2\phi + x_2 \cos^2\phi$, these integrals lead to

$$2\pi w_1 = \frac{1}{\sqrt{-\lambda x_3}} \left[2\phi + \frac{1}{2x_3}\left\{(x_1 + x_2)\phi + \frac{x_1 - x_2}{2}\sin^2\phi\right\}\right].$$

Writing in the values of x_1, x_2, x_3,

$$2\pi w_1 = 2\phi - \frac{\lambda}{2}\{-\lambda\alpha^2\phi + (\alpha + \dots)\sin^2\phi\}$$

$$= 2\phi - \frac{\lambda\alpha}{2}\sin^2\phi, \text{ neglecting } \lambda^2.$$

Therefore $\phi = \pi w_1$ approx., and using this we have

$$2\phi = 2\pi w_1 + \frac{\lambda\alpha}{2}\sin 2\pi w_1, \text{ approx.}$$

Now

$$x = x_1 \sin^2\phi + x_2\cos^2\phi$$

$$= \frac{x_1 + x_2}{2} + \frac{x_1 - x_2}{2}\cos 2\phi$$

$$= -\lambda\alpha^2 - \alpha\cos 2\phi, \text{ to this order,}$$

$$= -\lambda\alpha^2 - \alpha\cos\left[2\pi w_1 + \frac{\lambda\alpha}{2}\sin 2\pi w_1\right]$$

$$= -\lambda\alpha^2 + \alpha\left[\cos 2\pi w_1 - \sin 2\pi w_1\left(\frac{\lambda\alpha}{2}\sin 2\pi w_1\right)\right], \text{ up to } \lambda.$$

$$\therefore\ x = \alpha\cos 2\pi w_1 - \frac{\lambda\alpha^2}{4}(3 + \cos 4\pi w_1).$$

Thus x contains terms of the type $e^{\pm 2\pi i\tau w_1}$, where τ is 1 or 2, so that the quantum number n can, to this order, change by ± 1, ± 2, by the correspondence principle.

122. *Deductions from the theory.* If the smaller terms containing β, γ are omitted from the general formula (7) for the frequency ν, we have

$$\nu = \nu_0(n_1 - n_2) + \frac{h}{8\pi^2 A_0}(\pm 2m + 1), \text{ approximately} \dots (10).$$

In this result, the vibration term $\nu_0(n_1 - n_2)$ and the rotation term $\frac{h}{8\pi A_0}(\pm 2m + 1)$ occur independently. Thus for any given $n_1 - n_2$ there is a set of lines, given by the different values of m, forming a 'band,' and different values of $n_1 - n_2$ give different 'groups' of bands; the lines are not so deeply in the infra-red as for pure rotation spectra, as ν_0 is sufficient to make ν only a few μ's ($\mu = 10^{-4}$ cm.); for carbon monoxide, for instance, there is a band with its centre at 4·7 μ. The lines are spaced out at equal intervals $h^2/4\pi^2 A_0$ in frequency, but the central line corresponding to $m = 0$ is not present. This corresponds in absorption to the transition $0 \rightarrow 1$, so that $m = 0$ is not a possible initial state for absorption. The figure shows the position of the missing line or 'gap.'

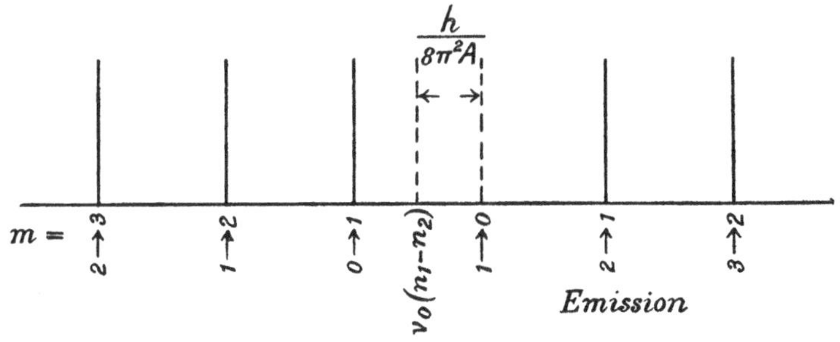

The second figure shows Imes' figure of the rotation-vibration band for HCl, with the gap at 3·48 μ.

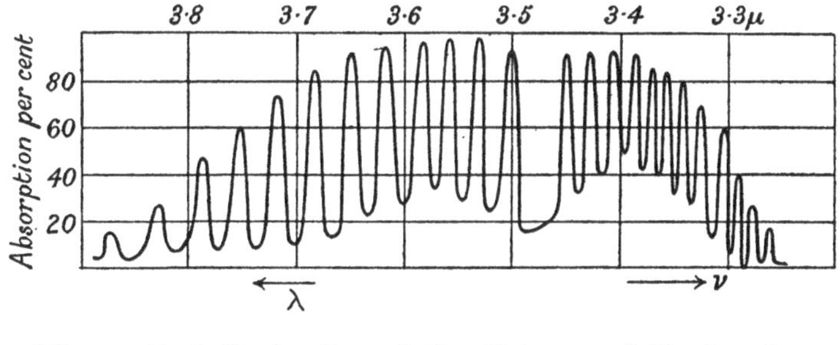

The gradual diminution of the distance of the bands as λ decreases is not in disaccord with their distances being equal for equal changes of ν, for $\nu = \frac{c}{\lambda}$. $\therefore\ |\Delta\nu| = \frac{c}{\lambda^2}|\Delta\lambda|$. Therefore for equal values of $|\Delta\nu|$, $|\Delta\lambda| \propto \lambda^2$ and $\therefore\ |\Delta\lambda|$ decreases as λ decreases.

For HCl are two known bands, with the gaps at 3·48 μ and

$1{\cdot}77\,\mu$, which have been observed by Imes[1]. These are absorption bands at ordinary temperatures and come out strongly, so that the initial states must have small energy of vibration. This corresponds to $n_2 = 0$ in the more general emission formula (7); the gap corresponding to $m = 0$ is given therefore by

$$\nu = \nu_0 n_1 + \beta h^2 n_1 + \gamma h n_1^2.$$

For
$$\left.\begin{aligned} n_1 &= 1, \quad \nu = \nu_0 + \gamma h + \beta h^2 \\ n_1 &= 2, \quad \nu = 2\nu_0 + 4\gamma h + 2\beta h^2 \end{aligned}\right\}.$$

Thus the fact that one ν is not exactly twice the other is accounted for by the higher approximation of formula (7).

123. *Effect of isotopes on the band spectrum.*

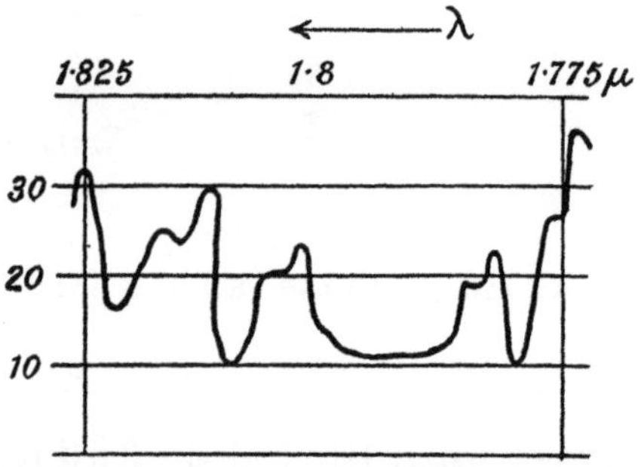

In Imes' figure for the $1{\cdot}77\,\mu$ band for HCl, certain subsidiary maxima appear in the graph of the absorption. Loomis[2] and also Kratzer[3] have explained these as being due to the isotopes Cl_{35}, Cl_{37} in the HCl molecule.

The position of the band $1{\cdot}77\,\mu$ for HCl is given by $n_2 = 0$, $n_1 = 2$, so that $\nu = 2\nu_0$ approx., and

$$\nu_0 = \frac{1}{2\pi}\sqrt{\frac{f''(a)}{M}},$$

where
$$\frac{1}{M} = \frac{1}{m_1} + \frac{1}{m_2}.$$

[1] E. S. Imes, Astrophys. Journ. 50, p. 251, 1919.
[2] F. W. Loomis, Astrophys. Journ. 52, p. 248, 1920.
[3] A. Kratzer, Zeitschr. für Physik, 3, p. 460, 1920.

For HCl_{35}, if m, $35m$ are the masses of the H and Cl_{35} atoms,

$$\frac{1}{M} = \frac{1}{m}\left(1 + \frac{1}{35}\right).$$

So for HCl_{37},
$$\frac{1}{M} = \frac{1}{m}\left(1 + \frac{1}{37}\right).$$

If ν_0 refers to the HCl_{35} band and $\nu_0 + \Delta\nu_0$ to the HCl_{37} band, then since $\nu_0{}^2 = C/M$, where C is $f''(a)/4\pi^2$,

$$\nu_0{}^2 = \frac{C}{m}\left(1 + \frac{1}{35}\right),$$

$$(\nu_0 + \Delta\nu_0)^2 = \frac{C}{m}\left(1 + \frac{1}{37}\right),$$

$$\therefore\ 2\nu_0\Delta\nu_0 = \frac{C}{m}\left(\frac{1}{37} - \frac{1}{35}\right),$$

$$\therefore\ \nu_0\Delta\nu_0 = -\frac{C}{m}\cdot\frac{1}{1295}.$$

On the left-hand side, the approximate ν_0 for either band may be used, or

$$\nu_0{}^2 = \frac{C}{m},$$

$$\therefore\ \frac{\Delta\nu_0}{\nu_0} = -\frac{1}{1295}.$$

Now
$$\nu = \frac{c}{\lambda},$$

$$\therefore\ \frac{\Delta\lambda_0}{\lambda_0} = -\frac{\Delta\nu_0}{\nu_0} = \frac{1}{1295},$$

and
$$\therefore\ \Delta\lambda_0 = \frac{1\cdot77}{1295}\mu = \frac{1\cdot77\times10^4}{1295}\text{Å}. = 13\cdot6\,\text{Å}.$$

Thus the maxima for the HCl_{37} band are slightly displaced from those of the HCl_{35} in the direction of greater wave length by about 13·6 Å.

Further, the atomic weight of Cl is 35·46 which indicates that it is a mixture of Cl_{37} and Cl_{35} in the proportions of about 1 : 3, so that in HCl there would be fewer HCl_{37} molecules than HCl_{35} ones; thus the intensities of the corresponding bands would be less for the former than the latter. This is all in agreement with Imes' figure, where the maxima of slightly greater wave length (corresponding to HCl_{37}) are not so high as those for HCl_{35}.

Further, Imes has found the value 14 Å. for the separation of the maxima which is in close agreement with the theory.

CHAPTER XVIII

BAND SPECTRA IN THE OPTICAL REGION; MOLECULAR ROTATION AND SPECIFIC HEAT

124. *Bands in the optical region.* When the temperature is sufficiently high to affect the electrons, they may be driven far enough from the axis of the molecule to have an appreciable angular momentum about the axis; the diatomic molecule would then correspond to a uniaxal body.

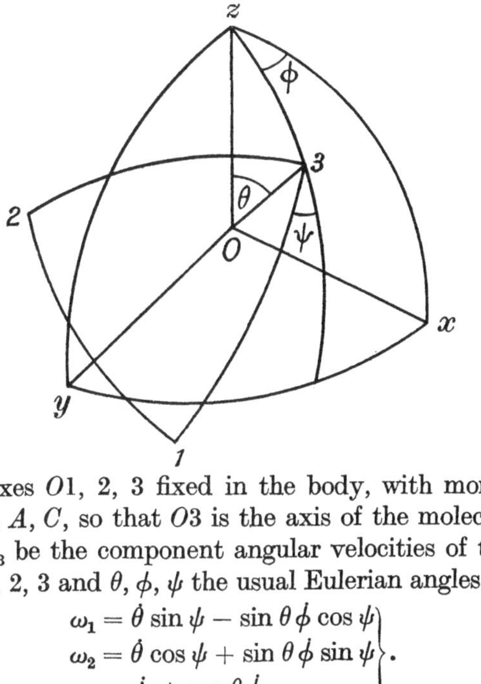

Take axes $O1$, 2, 3 fixed in the body, with moments of inertia A, A, C, so that $O3$ is the axis of the molecule. Let ω_1, ω_2, ω_3 be the component angular velocities of the body about $O1$, 2, 3 and θ, ϕ, ψ the usual Eulerian angles.

Then
$$\left.\begin{aligned}\omega_1 &= \dot{\theta}\sin\psi - \sin\theta\,\dot{\phi}\cos\psi\\ \omega_2 &= \dot{\theta}\cos\psi + \sin\theta\,\dot{\phi}\sin\psi\\ \omega_3 &= \dot{\psi} + \cos\theta\,\dot{\phi}\end{aligned}\right\}.$$

Therefore the kinetic energy

$$\begin{aligned}T &= \tfrac{1}{2}\left(A\omega_1^2 + A\omega_2^2 + C\omega_3^2\right)\\ &= \tfrac{1}{2}A\left(\dot{\theta}^2 + \sin^2\theta\,\dot{\phi}^2\right) + \tfrac{1}{2}C\left(\dot{\psi} + \cos\theta\,\dot{\phi}\right)^2.\end{aligned}$$

$$\left.\begin{aligned}\therefore\ p_1 &= \frac{\partial T}{\partial \dot\theta} = A\dot\theta\\ p_2 &= \frac{\partial T}{\partial \dot\phi} = A\sin^2\theta\dot\phi + C\cos\theta\,(\dot\psi + \cos\theta\dot\phi)\\ p_3 &= \frac{\partial T}{\partial \dot\psi} = C\,(\dot\psi + \cos\theta\,\dot\phi)\end{aligned}\right\}.$$

If Oz is chosen to be the 'invariable' line (the axis of the resultant angular momentum G, which is fixed in magnitude and direction in space), then

$$\left.\begin{aligned}-G\sin\theta\cos\psi &= A\omega_1 = A\,(\dot\theta\sin\psi - \sin\theta\cos\psi\dot\phi)\\ G\sin\theta\sin\psi &= A\omega_2 = A\,(\dot\theta\cos\psi + \sin\theta\sin\psi\dot\phi)\\ G\cos\theta &= C\omega_3 = C\,(\dot\psi + \cos\theta\dot\phi)\end{aligned}\right\}.$$

From the first two, solving for $\dot\theta$, $\dot\phi$, we have $\dot\theta = 0$ and $A\dot\phi = G$. Therefore θ is constant. Hence

$$p_1 = 0,\quad p_2 = A\sin^2\theta\dot\phi + p_3\cos\theta,$$

$$p_3 = C\,(\dot\psi + \cos\theta\dot\phi) = G\cos\theta,$$

$$\therefore\ p_1 = 0,\quad p_2 = G\sin^2\theta + G\cos^2\theta = G,\quad p_3 = G\cos\theta.$$

But $$H = T = \frac{1}{2}\left[\frac{p_1^2}{A} + \frac{(p_2 - p_3\cos\theta)^2}{A\sin^2\theta} + \frac{p_3^2}{C}\right],$$

$$\therefore\ H = \tfrac{1}{2}G^2\left(\frac{\sin^2\theta}{A} + \frac{\cos^2\theta}{C}\right) = \alpha_1 \quad \text{..................(1).}$$

The Hamiltonian equations,

$$\frac{\partial H}{\partial \phi} = \dot p_2,\quad \frac{\partial H}{\partial \psi} = \dot p_3,$$

yield $$p_2 = \text{constant} = \alpha_2,$$
$$p_3 = \text{constant} = \alpha_3.$$

The quantum conditions are

$$\oint p_2 d\phi = mh,\quad \oint p_3 d\psi = nh,$$

or $$2\pi\alpha_2 = mh,\quad 2\pi\alpha_3 = nh,$$

and $$\alpha_2 = G,\quad \alpha_3 = G\cos\theta,$$

so that $$G = \frac{mh}{2\pi},\quad \cos\theta = \frac{n}{m},$$

$$\therefore \quad a_1 = \frac{1}{2}\frac{m^2h^2}{4\pi^2}\left[\left(1 - \frac{n^2}{m^2}\right)\frac{1}{A} + \frac{n^2}{m^2}\cdot\frac{1}{C}\right].$$

Therefore the total energy in a stationary state is

$$E = \frac{h^2}{8\pi^2}\left\{\frac{m^2}{A} + n^2\left(\frac{1}{C} - \frac{1}{A}\right)\right\}.$$

125. If $C = 0$, then for a finite E, n must be 0, and we have

$$E = m^2h^2/8\pi^2A,$$

the formula of § 117.

Actually, when the electrons are excited, $C \neq 0$, and

$$E = E_e + h^2m^2/8\pi^2A,$$

where
$$E_e = \frac{h^2n^2}{8\pi^2}\left(\frac{1}{C} - \frac{1}{A}\right).$$

Electron changes will not much affect A, but may produce a small C so that E_e is large compared with $h^2m^2/8\pi^2A$.

In a transition where the frequency emitted is ν,

$$h\nu = E_1 - E_2 = (E_{e1} - E_{e2}) + h^2/8\pi^2A\,(m_1{}^2 - m_2{}^2).$$

Writing $E_{e1} - E_{e2} = h\nu_e$, where ν_e is the new frequency due to electron changes, $\nu = \nu_e + h/8\pi^2A\,(2m + 1)$, taking as before $m_1 \rightarrow m_2$ to be $m + 1 \rightarrow m$.

ν_e is large compared with $h/8\pi^2A$, so that the frequency ν, which in the absence of ν_e is in the infra-red, is by its presence lifted up into the visible and ultra-violet region.

The formula (8), § 120, now becomes

$$\nu = a \pm bm + cn^2,$$

where now

$$\left.\begin{aligned} a &= \nu_e + \frac{h}{8\pi^2A_0} + \nu_0(n_1 - n_2) + \beta h^2 n_1 + \gamma h(n_1{}^2 - n_2{}^2) \\ b &= \frac{h^2}{4\pi^2A_0} + \beta h^2 n_1 \\ c &= \beta h^2(n_1 - n_2) \end{aligned}\right\}.$$

126. *The band structure.* The figure shows the two parabolas $\nu = a \pm bm + cm^2$, with ν, m as abscissa and ordinate. The intersections of these with the lines $m = 0, 1, 2, \ldots$ give possible values of ν, which are projected on to the bands beneath which show the lines of the band in the two cases.

The line for which $m = 0$ is the 'null line' of the band and has zero intensity, producing in the upper band the 'gap' mentioned before; its position is located by the symmetry of the intensity of the lines on the two sides of it. The 'head' of a band is the end in the direction of low frequency.

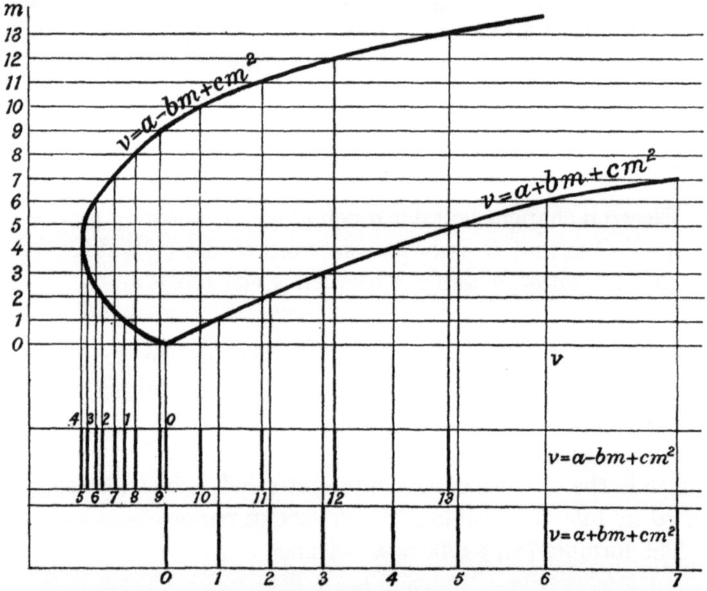

127. *Band systems.* The lines of a band are given by $\nu = a \pm bm + cm^2$, where m has integral values and a, b, c are the constants of the band. The frequency $\nu = a$, corresponding to $m = 0$, defines the 'null line' of the band.

The previous theory shows that a, b, c depend upon two quantum numbers n_1, n_2. A given difference $n_1 \sim n_2$ defines a '*group*' of bands and the groups form the band '*system*.'

A good example of a band system is that of the so-called 'cyanogen' bands due to the nitrogen molecule, shown in the figure at the top of the next page.

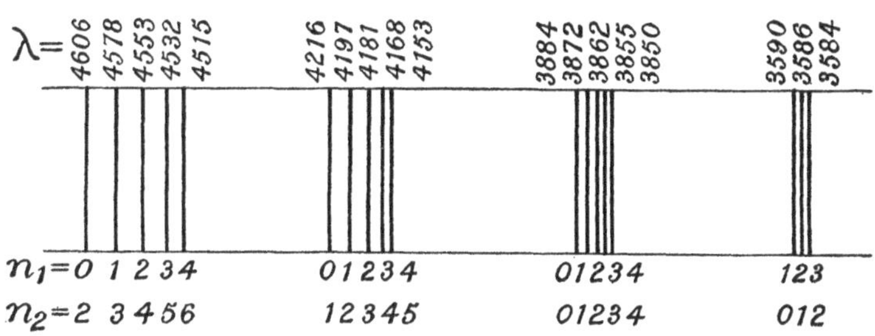

Each line in the above figure is the 'head' of a band (with its wave length), the band itself not being drawn. Four 'groups' of bands are indicated corresponding to $n_2 - n_1$ equal to 2, 1, 0, − 1.

The figure below is the actual appearance of the third 'group' of the above diagram, showing the bands and their constituent lines.

128. *Later theory.* Kramers and Pauli[1] and also Kratzer[2] have used as the dynamical model of the molecule a rigid body with a flywheel mounted inside it, the former to represent the nuclei and the latter the angular momentum of the electron system, and have given a theory of the 'gap' which occurs in the rotation-vibration spectrum.

Lately Fowler[3] has considered the intensities of the lines in a band by using the correspondence principle, and Nicholson[4]

[1] H. A. Kramers and W. Pauli, Jr, Zeitschr. für Physik, **13**, p. 351, 1923.

[2] A. Kratzer, Sitzungsber. Bayr. Akad. p. 107, 1922.

[3] R. H. Fowler, Phil. Mag. **49**, p. 1272, 1925.

[4] J. W. Nicholson, Phil. Mag. **50**, p. 650, 1925.

the general nature of band spectra in their relation to line spectra.

129. *The effect of molecular rotation on the specific heat of a diatomic gas.* The molecule has three degrees of translatory freedom and two effective ones of rotation, as rotation about the axis of symmetry could not be produced in encounters with other molecules. Each degree of freedom leads on the classical mechanics to a mean energy $\frac{1}{2}Rt$; the five degrees of freedom for the diatomic molecule lead to $\frac{5}{2}Rt$,

$$\therefore \quad E = \tfrac{5}{2}Rt, \text{ and } c_v = \frac{dE}{dt} = \tfrac{5}{2}R$$

determines the specific heat c_v from the energy E.

The experiments of Eucken[1] showed that the specific heat of hydrogen falls with falling temperature, and at 40° abs. reaches the value $3R/2$, the value for three degrees of freedom of the molecule. The hydrogen molecule then behaves as if it were monatomic; the rotational energy has vanished.

Ehrenfest[2] gave the theory of this effect, which follows. In the ordinary theory of gases[3] the probability of a state of energy E is proportional to $e^{-E/Rt}$, where t is the temperature.

Thus if $E_0, E_1, \ldots$ are the energies of different states of the system, the probabilities of these states are in the ratio

$$e^{-E_0/Rt} : e^{-E_1/Rt} : \ldots.$$

Therefore the mean value of the energy of a system whose possible energies are $E_0, E_1, E_2, \ldots$ is

$$\bar{E} = \frac{\sum_0^\infty E_n e^{-E_n/Rt}}{\sum_0^\infty e^{-E_n/Rt}} = \frac{\sum_0^\infty E_n e^{-\alpha E_n}}{\sum_0^\infty e^{-\alpha E_n}}, \text{ where } \alpha = \frac{1}{Rt},$$

$$= -\frac{\partial}{\partial \alpha} \log \left[\sum_0^\infty e^{-\alpha E_n}\right]$$

$$= -\frac{\partial}{\partial \alpha} \log U, \text{ where } U = \sum_0^\infty e^{-\alpha E_n}.$$

[1] A. Eucken, Sitzungsber. d. Preuss. Akad. p. 141, 1912.
[2] P. Ehrenfest, Verhandl. d. Deutsch. Phys. Ges. **15**, p. 451, 1913.
[3] J. H. Jeans, 'The dynamical theory of gases,' 3rd edition, 1920, or M. Planck, 'Wärmestrahlung,' p. 126, 5te Auflage, 1923.

For a rotating molecule (§ 117),

$$E_n = \frac{h^2n^2}{8\pi^2 A},$$

$$\therefore \quad U = \sum_0^\infty e^{-\frac{\alpha h^2 n^2}{8\pi^2 A}} = \sum_0^\infty e^{-\sigma n^2}, \text{ where } \sigma = \frac{\alpha h^2}{8\pi^2 A}.$$

Ehrenfest assumes the mean energy of a molecule to be twice this because there are *two* degrees of rotational freedom about an axis perpendicular to the axis of the molecule.

Therefore $\bar{E}$ is taken to be

$$-2\frac{\partial}{\partial \alpha}(\log U).$$

Therefore the specific heat c_v

$$= \frac{d\bar{E}}{dt} = -2\frac{\partial}{\partial t}\frac{\partial}{\partial \alpha}(\log U).$$

Since $Rt\alpha = 1$, this becomes

$$2R\alpha^2 \frac{\partial^2}{\partial \alpha^2}(\log U).$$

$$\therefore \quad c_v = 2R\alpha^2 \frac{\partial^2}{\partial \alpha^2}\log\left(\sum_0^\infty e^{-\sigma n^2}\right),$$

where $\sigma = h^2\alpha/8\pi^2 A.$

For very *low* temperatures, α is large and σ is too, therefore $e^{-\sigma}$ is very small and

$$\sum_0^\infty e^{-\sigma n^2} = 1 + e^{-\sigma} + e^{-4\sigma} + e^{-9\sigma} + \ldots$$

$$= 1 + e^{-\sigma} \text{ approx.},$$

and $\log\left(\sum_0^\infty e^{-\sigma n^2}\right) = \log(1 + e^{-\sigma}) = e^{-\sigma} \text{ approx.},$

$$\therefore \quad c_v \to 2R\alpha^2 \frac{\partial^2}{\partial \alpha^2}(e^{-h^2\alpha/8\pi^2 A})$$

$$\to 2R\alpha^2 \left(\frac{h^2}{8\pi^2 A}\right)^2 e^{-\sigma}$$

$$\to 2R\sigma^2 e^{-\sigma},$$

and this $\to 0 \text{ as } \sigma \to \infty.$

Thus the specific heat due to rotation vanishes at very low temperatures in agreement with Eucken's observation.

For very *high* temperatures, σ is small and

$$\sum_0^\infty e^{-\sigma m^2} \to \int_0^\infty e^{-\sigma m^2}\,dm = \frac{1}{2}\sqrt{\frac{\pi}{\sigma}},$$

$$\therefore\ c_v \to 2R\alpha^2 \frac{\partial^2}{\partial\alpha^2} \log\left(\frac{1}{2}\sqrt{\frac{\pi}{\sigma}}\right)$$

$$\to 2R\alpha^2 \frac{\partial^2}{\partial\alpha^2}[\text{constant} - \tfrac{1}{2}\log\sigma]$$

$$\to 2R\alpha^2 \frac{\partial^2}{\partial\alpha^2}[-\tfrac{1}{2}\log(h^2\alpha/8\pi^2 A)]$$

$$\to 2R\alpha^2[1/2\alpha^2]$$

$$\to R,$$

and has its full value for the two degrees of freedom of rotation.

CHAPTER XIX

THE KEPLERIAN ORBIT; THE DELAUNAY ELEMENTS

130. *The Keplerian orbit.* Using polar coordinates r, θ, ϕ,

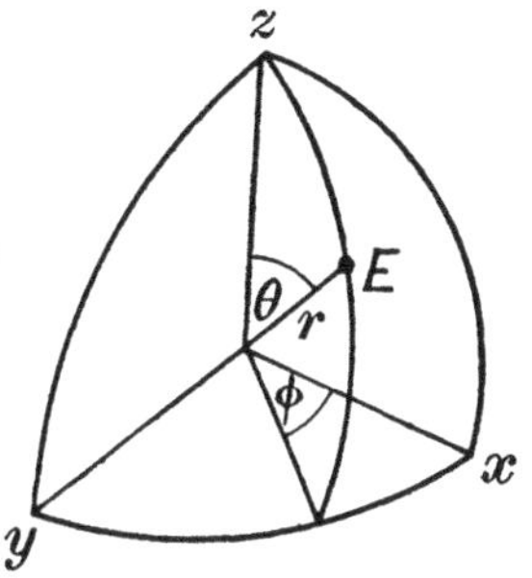

$$2T = m(\dot{r}^2 + r^2\dot{\theta}^2 + r^2\sin^2\theta\,\dot{\phi}^2),$$

$$V = -\frac{e^2}{r}.$$

$$\therefore\ H = \frac{m}{2}(\dot{r}^2 + r^2\dot{\theta}^2 + r^2\sin^2\theta\,\dot{\phi}^2) - \frac{e^2}{r},$$

and

$$p_1 = \frac{\partial T}{\partial \dot{r}} = m\dot{r}, \quad p_2 = \frac{\partial T}{\partial \dot{\theta}} = mr^2\dot{\theta},$$

$$p_3 = \frac{\partial T}{\partial \dot{\phi}} = mr^2\sin^2\theta\,\dot{\phi}.$$

$$\therefore\ H = \frac{1}{2m}\left(p_1^2 + \frac{1}{r^2}p_2^2 + \frac{1}{r^2\sin^2\theta}p_3^2\right) - \frac{e^2}{r} = \alpha_1.$$

The Hamilton-Jacobi equation is

$$\left(\frac{\partial S}{\partial r}\right)^2 + \frac{1}{r^2}\left(\frac{\partial S}{\partial \theta}\right)^2 + \frac{1}{r^2\sin^2\theta}\left(\frac{\partial S}{\partial \phi}\right)^2 - \frac{2me^2}{r} = 2m\alpha_1.$$

Since $\dfrac{\partial H}{\partial \phi} = \dot{p}_3$ and H does not contain ϕ, $\dot{p}_3 = 0$,

$$\therefore\ p_3 = \text{constant} = \alpha_3,$$

$$\therefore\ \frac{\partial S}{\partial \phi} = \alpha_3.$$

$$\therefore\ r^2\left[\left(\frac{\partial S}{\partial r}\right)^2 - \frac{2me^2}{r} - 2m\alpha_1\right] + \left[\left(\frac{\partial S}{\partial \theta}\right)^2 + \frac{\alpha_3^2}{\sin^2\theta}\right] = 0.$$

The variables separate. Writing

$$\left(\frac{\partial S}{\partial \theta}\right)^2 + \frac{\alpha_3^2}{\sin^2\theta} = \alpha_2^2,$$

we have
$$\left(\frac{\partial S}{\partial r}\right)^2 - \frac{2me^2}{r} - 2m\alpha_1 + \frac{\alpha_2^2}{r^2} = 0.$$

Thus
$$\left.\begin{aligned}\frac{\partial S}{\partial r} &= \left(2m\alpha_1 + \frac{2me^2}{r} - \frac{\alpha_2^2}{r^2}\right)^{\frac{1}{2}}\\ \frac{\partial S}{\partial \theta} &= \left(\alpha_2^2 - \frac{\alpha_3^2}{\sin^2\theta}\right)^{\frac{1}{2}}\\ \frac{\partial S}{\partial \phi} &= \alpha_3\end{aligned}\right\},$$

$$\therefore\ S = \int\left(\frac{\partial S}{\partial r}dr + \frac{\partial S}{\partial \theta}d\theta + \frac{\partial S}{\partial \phi}d\phi\right)$$

$$= \int dr\left(2m\alpha_1 + \frac{2me^2}{r} - \frac{\alpha_2^2}{r^2}\right)^{\frac{1}{2}} + \int d\theta\left(\alpha_2^2 - \frac{\alpha_3^2}{\sin^2\theta}\right)^{\frac{1}{2}} + \alpha_3\phi.$$

The orbit and the time are given by

$$\frac{\partial S}{\partial \alpha_1} = t + \beta_1,\quad \frac{\partial S}{\partial \alpha_2} = \beta_2,\quad \frac{\partial S}{\partial \alpha_3} = \beta_3 \quad \text{.........(1).}$$

The third of these gives

$$\phi - \int \frac{\alpha_3 d\theta}{\sin^2\theta\sqrt{\alpha_2^2 - \dfrac{\alpha_3^2}{\sin^2\theta}}} = \beta_3,$$

or
$$\phi - \beta_3 = \alpha_3 \int \frac{d\theta \operatorname{cosec}^2\theta}{\sqrt{\alpha_2^2 - \alpha_3^2(1 + \cot^2\theta)}}$$

$$= -\int \frac{\alpha_3 du}{\sqrt{\alpha_2^2 - \alpha_3^2 - \alpha_3^2 u^2}},$$

where $u = \cot\theta$,

$$= -\sin^{-1}\left(\frac{\alpha_3 \cot\theta}{\sqrt{\alpha_2^2 - \alpha_3^2}}\right).$$

Write
$$\cos i = \frac{\alpha_3}{\alpha_2}.$$

Then
$$\phi - \beta_3 = -\sin^{-1}(\cot\theta \cot i),$$

or
$$\cot\theta\cot i = \sin(\beta_3 - \phi),$$

$$\cos\theta\cos i = \sin\theta\sin i(\sin\beta_3\cos\phi - \cos\beta_3\sin\phi),$$

or if x, y, z are the Cartesian coordinates of the electron,

$$z\cos i = x\sin i\sin\beta_3 - y\sin i\cos\beta_3.$$

This gives the plane of the orbit. It shows that it makes an angle i with the plane $z = 0$, and also that its line of intersection with the plane $z = 0$ is $y = x \tan \beta_3$.

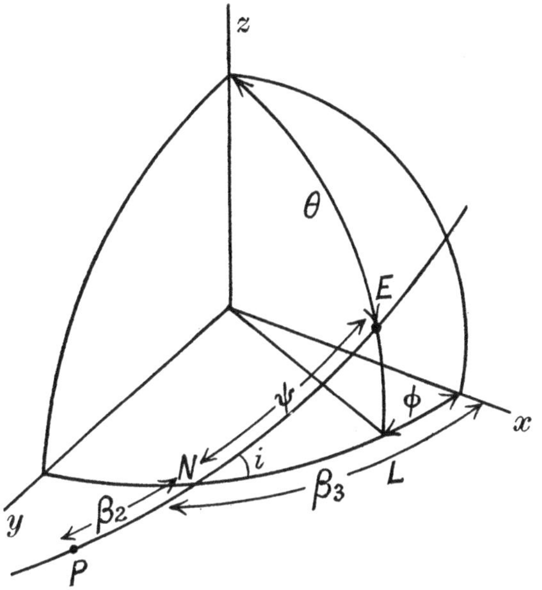

The second equation of (1) gives

$$\int -\frac{\alpha_2}{r^2} \frac{dr}{\sqrt{2m\alpha_1 + \frac{2me^2}{r} - \frac{\alpha_2{}^2}{r^2}}} + \alpha_2 \int \frac{d\theta}{\sqrt{\alpha_2{}^2 - \frac{\alpha_3{}^2}{\sin^2 \theta}}} = \beta_2,$$

or

$$\int \frac{\alpha_2 du}{\sqrt{-A + 2Bu - Cu^2}} - \alpha_2 \int \frac{d(\cos\theta)}{\sqrt{\alpha_2{}^2 - \alpha_3{}^2 - \alpha_2{}^2 \cos^2\theta}} = \beta_2,$$

where $\quad u = \frac{1}{r}, \quad A = -2m\alpha_1, \quad B = me^2, \quad C = \alpha_2{}^2.$

The first integral

$$= \alpha_2 \int \frac{du}{\sqrt{C(u_1 - u)(u - u_2)}},$$

where

$$u_1 + u_2 = 2B/C,$$
$$u_1 u_2 = A/C,$$

and writing $u = u_1 \sin^2 \chi + u_2 \cos^2 \chi$, it becomes

$$= 2\int d\chi = 2\chi,$$

since $$C = \alpha_2^2.$$

Hence we obtain

$$2\chi - \sin^{-1}\left(\frac{\alpha_2 \cos\theta}{\sqrt{\alpha_2^2 - \alpha_3^2}}\right) = \beta_2,$$

or $$2\chi - \sin^{-1}\left(\frac{\cos\theta}{\sin i}\right) = \beta_2,$$

using $$\cos i = \alpha_3/\alpha_2.$$

But from the spherical triangle NEL,

$$\sin EL = \sin i \sin NE,$$

or $$\cos\theta = \sin i \sin\psi,$$

denoting NE by ψ.

$$\therefore\ 2\chi - \psi = \beta_2,$$

or $$2\chi = \beta_2 + \psi.$$

But $$\frac{1}{r} = u = u_1 \sin^2 \chi + u_2 \cos^2 \chi$$

$$= \frac{u_1 + u_2}{2} + \frac{u_2 - u_1}{2} \cos 2\chi,$$

$$\therefore\ \frac{1}{r} = \frac{B}{C} + \frac{\sqrt{B^2 - AC}}{C} \cos(\psi + \beta_2),$$

or $$\frac{C}{B}\Big/ r = 1 + \sqrt{1 - \frac{AC}{B^2}} . \cos(\psi + \beta_2).$$

This is the polar equation of the elliptic orbit, and from it we see that β_2 is the distance from perihelion P to the node N.

If a, b, ϵ are the usual constants for the ellipse,

$$\frac{C}{B} = l = \frac{b^2}{a} \text{ and } \sqrt{1 - \frac{AC}{B^2}} = \epsilon,$$

$$\therefore\ \frac{b^2}{a^2} = 1 - \epsilon^2 = \frac{AC}{B^2},$$

$$\therefore\ a = \frac{B}{A},\quad b = \sqrt{\frac{C}{A}},$$

and $$A = -2m\alpha_1,\quad B = me^2,\quad C = \alpha_2^2.$$

131. *The action and angle variables.*

$$I_1 = \oint dr \frac{\partial S}{\partial r}$$

$$= \oint dr \sqrt{2m\alpha_1 + \frac{2me^2}{r} - \frac{\alpha_2{}^2}{r^2}} = \oint dr \sqrt{-A + \frac{2B}{r} - \frac{C}{r^2}}$$

$$= 2\pi\left(\frac{B}{\sqrt{A}} - \sqrt{C}\right). \qquad (\S\, 46)$$

$$\therefore\ I_1 = 2\pi\left(\frac{me^2}{\sqrt{-2m\alpha_1}} - \alpha_2\right),$$

$$I_2 = \oint d\theta \sqrt{\alpha_2{}^2 - \frac{\alpha_3{}^2}{\sin^2\theta}} = \oint d\theta\, \alpha_2 \sqrt{1 - \frac{\cos^2 i}{\sin^2\theta}}.$$

Now $\qquad \cot\theta \cot i = \cos(\phi + \beta_3) = \cos\phi',$

where $\qquad \phi' = \phi + \beta_3.$

$$\therefore\ I_2 = \int_0^{2\pi} \frac{\sin\phi' d\phi'}{\operatorname{cosec}^2\theta \cot i}\, \alpha_2 \sqrt{1 - \cos^2 i\left(1 + \frac{\cos^2\phi'}{\cot^2 i}\right)},$$

for in a libration of θ, ϕ' increases by 2π.

$$\therefore\ I_2 = \int_0^{2\pi} \frac{\alpha_2 \sin^2\phi' \sin i\, d\phi'}{\left(1 + \dfrac{\cos^2\phi'}{\cot^2 i}\right)\cot i} = \alpha_2 \int_0^{2\pi} \frac{\cos i \sin^2\phi' d\phi'}{(\cot^2 i + \cos^2\phi')}$$

$$= \alpha_3 \int_0^{2\pi} \frac{(1 - \cos^2\phi')\, d\phi'}{(\cot^2 i + \cos^2\phi')}, \text{ since } \alpha_2 \cos i = \alpha_3,$$

$$= \alpha_3 \int_0^{2\pi} \left(-1 + \frac{\operatorname{cosec}^2 i}{\cot^2 i + \cos^2\phi'}\right) d\phi'$$

$$= -2\pi\alpha_3 + 4\int_0^{\frac{\pi}{2}} \frac{\alpha_3 d\phi'}{\cos^2\phi' + \cos^2 i \sin^2\phi'}$$

$$= -2\pi\alpha_3 + \frac{2\pi\alpha_3}{\cos i},$$

$$\therefore\ I_2 = 2\pi(\alpha_2 - \alpha_3).$$

Also $\qquad I_3 = \oint \alpha_3 d\phi = 2\pi\alpha_3.$

From the above, $\alpha_3 = I_3/2\pi,$

$$\alpha_2 = (I_2 + I_3)/2\pi,$$

and
$$\frac{me^2}{\sqrt{-2m\alpha_1}} = (I_1 + I_2 + I_3)/2\pi,$$

or
$$\alpha_1 = -\frac{2\pi^2 me^4}{(I_1 + I_2 + I_3)^2}.$$

The corresponding frequencies satisfy

$$\omega_1 = \frac{\partial H}{\partial I_1} = \frac{\partial \alpha_1}{\partial I_1}, \text{ etc.}$$

$$\therefore\ \omega_1 = \omega_2 = \omega_3 = \frac{4\pi^2 me^4}{(I_1 + I_2 + I_3)^3} = \omega, \text{ say.}$$

The angle variables are

$$w_1 = \omega t + \delta_1,$$
$$w_2 = \omega t + \delta_2,$$
$$w_3 = \omega t + \delta_3,$$

and each coordinate is of the form

$$\Sigma C e^{2\pi i(\tau_1 w_1 + \tau_2 w_2 + \tau_3 w_3)},$$

or
$$\Sigma D e^{2\pi i \tau \omega t}.$$

and the motion is simply periodic of frequency ω.

132. *The Delaunay elements, J, v.*

Write
$$\left.\begin{aligned} I_1 + I_2 + I_3 &= J_1 \\ I_2 + I_3 &= J_2 \\ I_3 &= J_3 \end{aligned}\right\}.$$

The corresponding angle variables v_1, v_2, v_3 are therefore (cf. § 56) given by

$$w_1 = v_1, \qquad \text{or} \qquad v_1 = w_1,$$
$$w_2 = v_1 + v_2, \qquad\qquad v_2 = w_2 - w_1,$$
$$w_3 = v_1 + v_2 + v_3, \qquad\qquad v_3 = w_3 - w_2,$$

or
$$v_1 = \omega t + \delta_1,$$
$$v_2 = \delta_2 - \delta_1,$$
$$v_3 = \delta_3 - \delta_2,$$

so that v_2, v_3 are constants.

In terms of the new quantities,

$$\alpha_1 = \frac{-2\pi^2 me^4}{J_1^2}, \quad \alpha_2 = \frac{J_2}{2\pi}, \quad \alpha_3 = \frac{J_3}{2\pi}.$$

Also

$$a = \frac{B}{A} = \frac{e^2}{-2\alpha_1} = \frac{J_1^2}{4\pi^2 me^2},$$

$$b = \sqrt{\frac{C}{A}} = \frac{\alpha_2}{\sqrt{-2m\alpha_1}} = \frac{J_1 J_2}{4\pi^2 me^2},$$

$$\frac{b}{a} = \frac{J_2}{J_1},$$

$$\epsilon = \sqrt{1 - \frac{b^2}{a^2}} = \sqrt{1 - \frac{J_2^2}{J_1^2}},$$

$$\cos i = \frac{\alpha_3}{\alpha_2} = \frac{J_3}{J_2}.$$

Expressing S in terms of the new quantities,

$$S = \int dr \left(\frac{-4\pi^2 m^2 e^4}{J_1^2} + \frac{2me^2}{r} - \frac{J_2^2}{4\pi^2 r^2}\right)^{\frac{1}{2}} + \int \frac{d\theta}{2\pi}\left(J_2^2 - \frac{J_3^2}{\sin^2\theta}\right)^{\frac{1}{2}} + \frac{J_3\phi}{2\pi},$$

$$\left.\begin{aligned}
\therefore\ v_1 &= \frac{\partial S}{\partial J_1} = \int \frac{4\pi^2 m^2 e^4}{J_1^3}\left(\frac{-4\pi^2 m^2 e^4}{J_1^2} + \frac{2me^2}{r} - \frac{J_2^2}{4\pi^2 r^2}\right)^{-\frac{1}{2}} dr \\
v_2 &= \frac{\partial S}{\partial J_2} = \int \frac{-J_2}{4\pi^2 r^2}\left(\frac{-4\pi^2 m^2 e^4}{J_1^2} + \frac{2me^2}{r} - \frac{J_2^2}{4\pi^2 r^2}\right)^{-\frac{1}{2}} dr \\
&\qquad + \int \frac{J_2}{2\pi}\, d\theta \left(J_2^2 - \frac{J_3^2}{\sin^2\theta}\right)^{-\frac{1}{2}} \\
v_3 &= \int \frac{-J_3 d\theta}{2\pi \sin^2\theta}\left(J_2^2 - \frac{J_3^2}{\sin^2\theta}\right)^{\frac{1}{2}} + \frac{\phi}{2\pi}
\end{aligned}\right\}.$$

Then using the integrations worked out above (pp. 188–190), we have

$$\left.\begin{aligned}
2\pi v_1 &= f(r) \\
2\pi v_2 &= 2\chi - \psi = \beta_2 \\
2\pi v_3 &= (\beta_3 - \phi) + \phi = \beta_3
\end{aligned}\right\},$$

$v_1 = \omega t + \delta_1$, so that $2\pi v_1$ is the 'mean anomaly' of astronomy and the first equation expresses r in terms of t.

The second and third show that $2\pi v_2$ is the distance from perihelion to the node and that $2\pi v_3$ is the azimuth of the node itself (figure on p. 189).

These quantities $J_1, J_2, J_3, v_1, v_2, v_3$ are the '*elements' of Delaunay* used in astronomical theory; they have especially been used as the starting point by Burgers[1] in working out the perturbations of the Keplerian orbit due to electric and magnetic fields. He introduces them by the use of a contact transformation which arises in the reduction of the equations of motion to a Hamiltonian system of the sixth order in the problem of three bodies[2].

[1] J. M. Burgers, 'Het Atoommodel van Rutherford-Bohr,' Archives du Musée Seyler, III, 4, p. 80 et seq. Haarlem, 1919.

[2] E. T. Whittaker, 'Analytical Dynamics,' p. 349, Cambridge, 1917.

CHAPTER XX

THE ASTRONOMICAL THEORY OF PERTURBATIONS AND ATOM MECHANICS

133. In the theory of the Stark effect (ch. IX) it was possible to calculate the effect of a strong electric field on the orbit of the electron by the procedure of 'separation of variables'; the relativity effect could be neglected as it was small compared with that of the field.

If, however, the electric field is not a strong one, so that the relativity effect is no longer negligible, no set of generalised coordinates can be found for which a separation of variables is possible.

This latter problem has however been solved[1] by regarding the field as producing a 'perturbation' of the normal Keplerian orbit of the electron and applying to the atomic system the theory of perturbations of celestial mechanics.

The introduction of the perturbation theory of astronomy into atom mechanics is due to Bohr[2], who first considered the perturbations of a periodic orbit.

Bohr's theory of a perturbed periodic system. The theory will perhaps best be understood if a definite case, such as the Keplerian orbit, is kept in mind; here there are three coordinates q_1, q_2, q_3 for the electron, and the motion is simply periodic.

Suppose the undisturbed motion found from the Hamilton-Jacobi equation and given by

$$\frac{\partial S}{\partial \alpha_1} = t + \beta_1, \quad \frac{\partial S}{\partial \alpha_2} = \beta_2, \quad \frac{\partial S}{\partial \alpha_3} = \beta_3 \quad \text{.........(1)},$$

so that the orbit is characterised by the constants

$$\alpha_1, \alpha_2, \alpha_3, \beta_1, \beta_2, \beta_3.$$

[1] H. A. KRAMERS, Zeitschr. für Physik, 3, p. 199, 1920.
[2] N. BOHR, 'Q.L.S.' Part II, pp. 41–98.

Let the system be subject to some small external field of force so that the motion is no longer simply periodic. We define 'the osculating orbit' at any given instant as the orbit which would result if the external forces suddenly vanished at that moment; the parameters $\alpha_1, \alpha_2, \alpha_3, \beta_1, \beta_2, \beta_3$ which define the osculating orbit now vary slowly with the time.

Let Ω be the potential energy due to the external field expressed in terms of q_1, q_2, q_3. Since by solving (1), the q's can be expressed in terms of the α's, β's and t, Ω is known in terms of $\alpha_1, \alpha_2, \alpha_3, \beta_1, \beta_2, \beta_3, t$.

From the perturbation theory of celestial mechanics (§ 47)

$$\dot{\alpha} = -\frac{\partial\Omega}{\partial\beta}, \quad \dot{\beta} = \frac{\partial\Omega}{\partial\alpha}, \text{ for each } \alpha \text{ and } \beta \quad \text{......(2).}$$

These equations give the perturbing effect of the field on the orbit at any instant.

But a detailed examination from instant to instant of the perturbations is not required, but only the 'secular perturbations,' which are the total variations of the α's and the β's over a time large compared with the period of the osculating orbit.

Let λ be a small quantity of the same order as the ratio of the perturbing force to the force on the electron due to the nucleus, and let terms of order λ^2 be neglected.

The constants α_1, β_1 differ essentially from the other α's and β's as is seen by the different form of the equation they satisfy in (1).

α_1 is the total energy, and since by the conservation of energy $\alpha_1 + \Omega$ is constant during the motion, the changes in α_1 can only be of order λ even after a long period of time. Also β_1 is a special value of the time at which the electron passes some given point of the orbit, and in discussing secular perturbations the origin of measurement of time is immaterial. Also σ, the period of the undisturbed orbit, depends upon α_1 only (in the Keplerian orbit $\sigma^2 = 8\alpha_1^3/me^4$, § 32), so that its secular changes are only of order λ. Hence in finding the mean value of Ω over a period of the osculating orbit, we may take it for a period of the undisturbed orbit, because the change of σ is of order λ, and if this is taken into account, the effect on Ω, itself of order λ, is only of order λ^2.

But for α_2, α_3, we have

$$\dot{\alpha} = -\frac{\partial \Omega}{\partial \beta},$$

$$\therefore\ (\alpha)_{t_1}^{t_2} = -\int_{t_1}^{t_2} \frac{\partial \Omega}{\partial \beta}\, dt = -\left(\overline{\frac{\partial \Omega}{\partial \beta}}\right)(t_2 - t_1) \quad \ldots\ldots(3),$$

where the bar denotes the mean value in the interval t_1 to t_2.

Now $\overline{\dfrac{\partial \Omega}{\partial \beta}}$ is of order λ, and if the interval $t_2 - t_1$ is large, of order σ/λ (i.e. contains a large number of orbital periods), the right-hand side of (3) is finite; so that the change of α in a time large compared with σ is of the same order as α, for each of α_2, α_3. So for β_2, β_3. These are the 'secular perturbations' of the orbital constants.

If the mean value of Ω is denoted by ψ, so that $\psi = \overline{\Omega}$, then equations (2) become, if the mean values are taken over a large number of periods,

$$\left.\begin{aligned} \bar{\dot{\alpha}} &= -\frac{\partial \overline{\Omega}}{\partial \beta} = -\frac{\partial \psi}{\partial \beta} \\ \bar{\dot{\beta}} &= +\frac{\partial \psi}{\partial \alpha} \end{aligned}\right\}.$$

Denoting the mean value of $\dfrac{d\alpha}{dt}$ by $\dfrac{D\alpha}{Dt}$, so that $\bar{\dot{\alpha}} = \dfrac{D\alpha}{Dt}$, we have

$$\frac{D\alpha}{Dt} = -\frac{\partial \psi}{\partial \beta}, \quad \frac{D\beta}{Dt} = \frac{\partial \psi}{\partial \alpha} \quad \ldots\ldots\ldots\ldots\ldots\ldots(4),$$

where ψ is the mean value of the disturbing potential Ω.

Now consider the effect on the total energy of the system due to the slow *uniform* establishment of the field Ω. Let the change occur in an interval 0 to θ, where θ is long and of order σ/λ.

$\Delta\alpha_1$ due to the field = work done by the external forces from $t = 0$ to $t = \theta$, and this

$$= -\int_0^\theta \left(\frac{t}{\theta} d\Omega\right) = \left(-\frac{t\Omega}{\theta}\right)_0^\theta + \int_0^\theta \frac{\Omega}{\theta} dt$$

$$= -\Omega_\theta + \psi,$$

for the second term is the mean value of Ω over the long period.

$$\therefore \; \Delta\alpha_1 + \Omega_\theta = \psi \text{ ....(5).}$$

Therefore the change of the total energy due to the slow and uniform establishment of the field is just equal to this function ψ.

Thus the problem is reduced to the solution of the Hamilton equations for a system of only *two* degrees of freedom, with ψ as the Hamiltonian function; for equations (4) are of the canonical form, and by equation (5) ψ is the total energy of the perturbations.

The stationary states are then determined by the use of new action variables $\mathcal{J}_2$, $\mathcal{J}_3$, where $\mathcal{J}_2 = \oint \alpha_2 d\beta_2$, $\mathcal{J}_3 = \oint \alpha_3 d\beta_3$. This theory is carried out for the Stark effect in § 138.

Bohr, in his theory of the Stark effect which follows, replaces the procedure of this last paragraph by a direct solution of the perturbation equations, which discloses a new frequency ω_H; he then determines the stationary states by the condition $d\psi = \omega_H dI_H$, which assumes[1] that the frequency of the slow variation of the orbit has a relation to the additional energy ψ due to the external field of the same kind as the relation (expressed by $dE = \omega dI$) between the frequency and energy of the original periodic orbit.

134. *Bohr's theory of the Stark effect.* O is the nucleus, Oz the direction of the external electric field F, $O\xi$ the major axis of the orbit, $O\eta$ at right angles to it in the plane of the orbit, OP is the line from O to the electron. These lines meet

[1] See N. Bohr, 'Q.L.S.' Part I, p. 23; Proc. Phys. Soc. London, **35**, p. 290, 1923.

a unit sphere centre O in the points z, ξ, η, P. The polar coordinates of the electron in the plane of its orbit are r, θ, so that $\xi = r\cos\theta$, $\eta = r\sin\theta$.

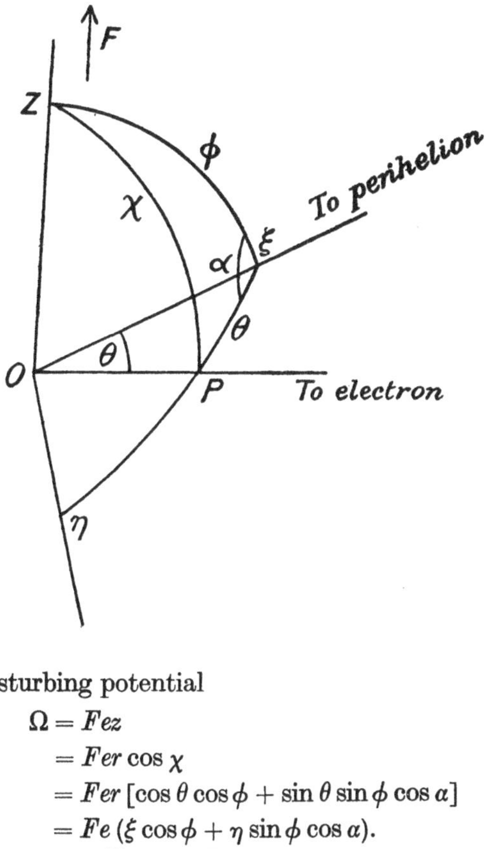

The disturbing potential

$$\begin{aligned}\Omega &= Fez\\ &= Fer\cos\chi\\ &= Fer\,[\cos\theta\cos\phi + \sin\theta\sin\phi\cos\alpha]\\ &= Fe\,(\xi\cos\phi + \eta\sin\phi\cos\alpha).\end{aligned}$$

$$\therefore\quad \psi = \bar{\Omega} = Fe\,(\bar{\xi}\cos\phi + \bar{\eta}\sin\phi\cos\alpha).$$

$\bar{\xi}$, $\bar{\eta}$ is called the 'electric centre' of the orbit.

If u is the eccentric angle, it is known from the theory of the Keplerian ellipse that $nt = u - \epsilon\sin u$, where $\frac{2\pi}{n}$ is the period τ, and ϵ is the eccentricity.

Hence $\bar{\xi} = \frac{1}{\tau}\int_0^\tau \xi\,dt = \frac{1}{\tau}\int_0^\tau (a\cos u - a\epsilon)\,dt$

$$= \frac{1}{\tau}\int_0^{2\pi} a(\cos u - \epsilon)(1 - \epsilon\cos u)\frac{du}{n}$$

$$= \frac{a}{2\pi}[-2\pi\epsilon - \pi\epsilon]$$

$$= -\frac{3a\epsilon}{2}.$$

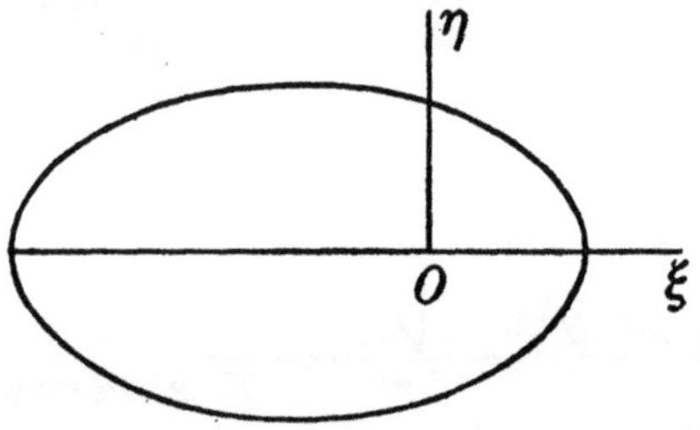

So $\bar{\eta} = \frac{1}{\tau}\int_0^\tau (b\sin u)\,dt = \frac{1}{\tau}\int_0^{2\pi} (b\sin u)(1 - \epsilon\cos u)\frac{du}{n}$

$$= 0,$$

$$\therefore\ \psi = -\tfrac{3}{2}Fea\epsilon\cos\phi.$$

Since ψ, the total additional energy due to the perturbations, is constant throughout the motion,

$$(-\tfrac{3}{2}a\epsilon)\cos\phi = \frac{\psi}{Fe} = \text{constant}.$$

Therefore the projection of the electric centre on Oz is a fixed point, or the electric centre moves in a fixed plane at right angles to the field.

To find the motion of the electric centre let its coordinates be x, y, z, so that $x^2 + y^2 + z^2 = (\tfrac{3}{2}a\epsilon)^2$.

The angular momentum of the electron is mH, where H is the usual h of elliptic motion, so that

$$H = \sqrt{\mu l} = \sqrt{\frac{e^2}{m}a(1 - \epsilon^2)}.$$

If λ, μ, ν are the components of this angular momentum

$$\lambda^2 + \mu^2 + \nu^2 = m^2 H^2 = e^2 ma\,(1 - \epsilon^2).$$

Eliminating ϵ,

$$x^2 + y^2 + z^2 = \tfrac{9}{4}a^2 \left\{1 - \frac{\lambda^2 + \mu^2 + \nu^2}{mae^2}\right\} \quad \text{.........(1).}$$

Also the resultant angular momentum is perpendicular to the plane of the orbit, so that

$$x\lambda + y\mu + z\nu = 0 \quad \text{..................(2).}$$

Also z has just been shown to be constant.

If ξ, η, ζ is the electron, its equations of motion are

$$\dot{\lambda} = eF\eta, \quad \dot{\mu} = -\,eF\xi, \quad \dot{\nu} = 0,$$

or taking mean values

$$\frac{D\lambda}{Dt} = eF\bar{\eta} = eFy, \quad \frac{D\mu}{Dt} = -\,eF\bar{\xi} = eFx,$$

and ν is constant.

Using dashes to denote $\dfrac{D}{Dt}$, we have

$$\lambda' = eFy, \quad \mu' = -\,eFx \quad \text{..............(3).}$$

Also differentiating (1),

$$xx' + yy' = -\,\frac{9a}{4me^2}(\lambda\lambda' + \mu\mu'), \text{ since } z \text{ and } \nu \text{ are constant,}$$

$$= -\,\frac{9aF}{4me}(\lambda y - \mu x), \text{ using (3)} \quad \text{..................(4).}$$

Differentiating (2),

$$\lambda x' + \mu y' = -\,(x\lambda' - y\mu')$$

$$= 0, \text{ using (3)} \quad \text{..................(5).}$$

Therefore solving (4) and (5) for x' and y' we find

$$\frac{x'}{\mu} = \frac{y'}{-\,\lambda} = \frac{9aF}{4me},$$

$$\therefore\ x'' = \frac{9aF}{4me}\mu' = -\,\frac{9aF^2}{4m}x.$$

So
$$y'' = -\frac{9aF^2}{4m}\,.\,y,$$

$$\therefore\ \left.\begin{aligned} x &= A\cos(2\pi\omega_F t + \alpha)\\ y &= B\cos(2\pi\omega_F t + \beta)\end{aligned}\right\} \quad\ldots\ldots\ldots\ldots\ldots(6),$$

where
$$(2\pi\omega_F)^2 = \frac{9aF^2}{4m}.$$

But for the Keplerian orbit

$$a = \frac{I^2}{4\pi^2 me^2}, \qquad (\S\ 32)$$

$$\therefore\ (2\pi\omega_F)^2 = \frac{9I^2F^2}{16\pi^2m^2e^2},$$

$$\therefore\ \omega_F = \frac{3IF}{8\pi^2 me} \quad\ldots\ldots\ldots\ldots\ldots\ldots\ldots(7).$$

Equations (6) show that the electric centre describes an ellipse with its *centre* on the axis of z, the frequency in the ellipse being ω_F, given by (7).

Thus the effect of the field is to introduce a new frequency into the system, which has now a doubly periodic motion.

The stationary states are fixed by two quantum conditions $I = nh$, $I_F = n_F h$, where I, I_F satisfy

$$\delta E = \omega\delta I + \omega_F\delta I_F \text{ and } \omega I + \omega_F I_F = \bar{A}. \quad (\S\ 53)$$

The special case where the original orbit is a circle whose plane is perpendicular to F indicates that the dependence of ω on I is that holding in a simple Keplerian orbit, so that

$$\delta E = +\frac{4\pi^2e^4m}{I^3}\delta I + \frac{3IF}{8\pi^2me}\delta I_F,$$

whence
$$E = -\frac{2\pi^2me^4}{I^2} + \frac{3II_FF}{8\pi^2me}.$$

Using $I = nh$, $I_F = n_F h$, the energy of the stationary states is given by

$$E = -\frac{2\pi^2me^4}{h^2n^2} + \frac{3h^2F}{8\pi^2me}nn_F \quad\ldots\ldots\ldots\ldots(8),$$

the formula found in § 71.

135. Further, $z = -\frac{3}{2}a\epsilon\cos\phi$, so that z_m, the maximum value of z, is $\frac{3a}{2}$, in the limiting case where the orbit degenerates into a straight line along Oz; this limit is on physical grounds ruled out, so that z is always $< z_m$ and never equal to it.

But
$$a = \frac{I^2}{4\pi^2 me^2},$$

so that
$$z_m = \frac{3I^2}{8\pi^2 me^2} = \frac{3h^2n^2}{8\pi^2 me^2}.$$

But $\psi = Fez$ and ψ is the additional energy due to the perturbations, which from (8) is $\frac{3h^2Fnn_F}{8\pi^2 me}$,

$$\therefore \quad z = \frac{3h^2nn_F}{8\pi^2 me^2},$$

$$\therefore \quad z/z_m = n_F/n.$$

Since $z < z_m$ and never equal to it, n_F may be any integer *less* than n (cf. § 73).

136. *The polarisation.* Consider the motion of the electron referred to axes x, y, z rotating round Oz with the electric centre, that is with frequency ω_F.

The electron has a frequency ω_1 in its orbit which itself goes through a cycle of changes of shape and position *twice* in each revolution of the electric centre, so that the frequency of this cycle is $2\omega_F$ ($\equiv \omega_2$, suppose).

[The path of the electric centre is an ellipse with its *centre* on Oz, so that distances from Oz repeat themselves twice in a revolution.]

Thus the coordinates x, y, z of the electron referred to these axes are doubly periodic with frequencies of type $\tau_1\omega_1 + \tau_2\omega_2$.

Referred to fixed axes X, Y, Z, where OZ coincides with Oz, the Z coordinate is still of the type with frequencies $\tau_1\omega_1 + \tau_2\omega_2$, but the X, Y coordinates have frequencies $\tau_1\omega_1 + \tau_2\omega_2 \pm \omega_F$ (cf. § 79).

Writing $\omega = \omega_1 \pm \omega_F$, so that ω is the frequency of the electron round OZ, the Z frequencies are of the type

$$\tau_1 (\omega \mp \omega_F) + \tau_2 \,.\, 2\omega_F, \text{ since } \omega_2 = 2\omega_F,$$

and the X, Y frequencies are of the type

$$\tau_1 (\omega \mp \omega_F) + \tau_2 \,.\, 2\omega_F \pm \omega_F,$$

or the Z frequencies are of the type

$$\tau\omega + \tau_F \omega_F,$$

where $$\tau + \tau_F \text{ is } \tau_1 \pm \tau_1 + 2\tau_2$$

and is therefore *even*, and the X, Y frequencies are of the type

$$\tau\omega + \tau_F \omega_F,$$

where $$\tau + \tau_F \text{ is } \tau_1 \pm \tau_1 + 2\tau_2 \pm 1$$

and is therefore *odd*.

These are the polarisation conditions found in § 72.

137. *Asymmetry in the Stark effect.* Stark[1] has observed that while under usual conditions the components of each hydrogen line exhibit complete symmetry with respect to the original line, a striking asymmetry occurs when the spectrum is excited by electron bombardment. If the electrons are moving in the same direction as the electric force used to produce the Stark effect, the components on the long-wave side of the original line are much more intense than those on the short-wave side. The equation $z/z_m = n_F/n$ for the position of the plane in which the electric centre moves, shows the dependence of a given line upon the position of this plane. The probability of this plane being displaced in the direction of the incident electrons is greater than of its being displaced in the opposite direction, so that more changes of n_F are probable in the positive than in the negative sense, and therefore a greater intensity on the one side of the original line than on the other[2].

[1] J. Stark, Ann. d. Physik, **56**, p. 569, 1918.
[2] N. Bohr, Phil. Mag. **30**, p. 404, 1915.

138. *Use of the Delaunay elements of the orbit to determine the Stark effect.* Here the J_2, J_3, v_2, v_3 of the undisturbed Keplerian orbit (§ 132) correspond to the $\alpha_2, \alpha_3, \beta_2, \beta_3$ of Bohr's theory (§ 133).

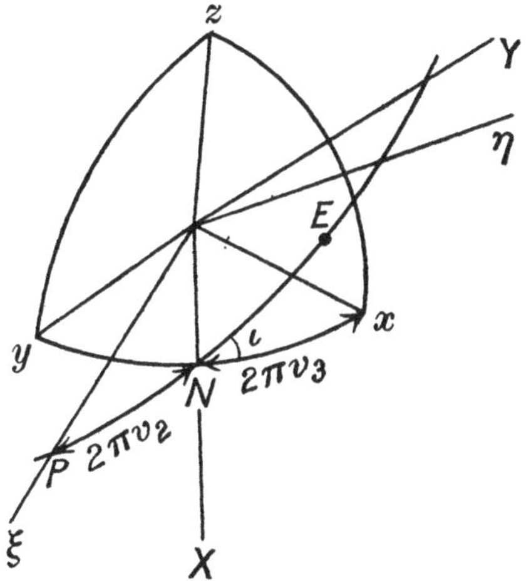

The disturbing potential

$$\Omega = Fez = FeY \sin i,$$

and $$Y = \eta \cos 2\pi v_2 - \xi \sin 2\pi v_2,$$

by transformation from the axes X, Y to the axes ξ, η.

$$\therefore \quad \Omega = Fe\,[\eta \cos 2\pi v_2 - \xi \sin 2\pi v_2] \sin i,$$

$$\therefore \quad \psi = \overline{\Omega} = Fe\,[\bar{\eta} \cos 2\pi v_2 - \bar{\xi} \sin 2\pi v_2] \sin i.$$

But $$\bar{\xi} = -\tfrac{3}{2} a\epsilon, \quad \bar{\eta} = 0, \qquad (\S\ 134)$$

$$\therefore \quad \psi = Fe\left(\frac{3a\epsilon}{2}\right) \sin 2\pi v_2 \sin i.$$

Using the orbital data of § 132,

$$\psi = \tfrac{3}{2} Fe \sin 2\pi v_2 \frac{J_1^2}{4\pi^2 me^2} \sqrt{\left(1 - \frac{J_2^2}{J_1^2}\right)\left(1 - \frac{J_3^2}{J_2^2}\right)} \ldots (1).$$

The equations $$\frac{D\alpha}{Dt} = -\frac{\partial\psi}{\partial\beta}, \quad \frac{D\beta}{Dt} = \frac{\partial\psi}{\partial\alpha}$$

of the theory, which give the secular perturbations of α_2, α_3, β_2, β_3, become here

$$\frac{DJ}{Dt} = -\frac{\partial\psi}{\partial v} \text{ and } \frac{Dv}{Dt} = \frac{\partial\psi}{\partial J}, \text{ for } J_2, J_3, v_2, v_3.$$

$$\therefore \quad \frac{DJ_3}{Dt} = -\frac{\partial\psi}{\partial v_3} = 0,$$

so that J_3 has no secular change; the general theory showed that α_1, here J_1, has no secular change, so that if $\mathscr{J}_1$, $\mathscr{J}_2$, $\mathscr{J}_3$ are the values of J_1, J_2, J_3 for the perturbed orbit, $\mathscr{J}_1 = J_1$ and $\mathscr{J}_3 = J_3$.

Now $$\mathscr{J}_2 = \oint \alpha_2 d\beta_2, \text{ in the theory (§ 133),}$$

$$= \oint J_2 dv_2 \text{ here.}$$

Writing $$\left(\frac{8\pi^2 me}{3FJ_1^2}\psi\right)^2 = B,$$

equation (1) gives

$$\sin^2 2\pi v_2 = \frac{B}{\left(1 - \frac{J_2^2}{J_1^2}\right)\left(1 - \frac{J_3^2}{J_2^2}\right)},$$

or $$\sin^2 2\pi v_2 = \frac{ABx}{(A - x)(x - C)},$$

where $$A = J_1^2, \quad x = J_2^2, \quad C = J_3^2.$$

Taking logarithms and differentiating,

$$4\pi \cot 2\pi v_2 \frac{dv_2}{dx} = \frac{1}{x} + \frac{1}{A - x} - \frac{1}{x - C} = \frac{x^2 - AC}{x(A - x)(x - C)},$$

$$\therefore \quad 4\pi \frac{dv_2}{dx} = \frac{x^2 - AC}{x(A - x)(x - C)} \frac{\sqrt{ABx}}{\sqrt{(A - x)(x - C) - ABx}}.$$

$$\therefore \quad \mathscr{J}_2 = \oint J_2 dv_2$$

$$= \oint \frac{\sqrt{AB}(x^2 - AC)\,dx}{4\pi(A - x)(x - C)\sqrt{(A - x)(x - C) - ABx}}$$

$$= \oint \frac{\sqrt{AB}(x^2 - AC)\,dx}{4\pi(A - x)(x - C)\sqrt{(\alpha - x)(x - \beta)}},$$

where $\alpha + \beta = A + C - AB, \quad \alpha\beta = AC$

$$= 2\int_\beta^\alpha \frac{\sqrt{AB}\,(x^2 - AC)\,dx}{4\pi\,(A-x)\,(x-C)\sqrt{(\alpha - x)\,(x-\beta)}}.$$

Writing $x = \alpha \sin^2\psi + \beta\cos^2\psi$, the integral is equal to

$$\frac{\sqrt{AB}}{2\pi}\int_0^{\frac{\pi}{2}} \frac{(x^2 - AC)}{(A-x)\,(x-C)}\,(2d\psi)$$

$$= \frac{\sqrt{AB}}{\pi}\int_0^{\frac{\pi}{2}} d\psi \left[-1 + \frac{A}{A-x} - \frac{C}{x-C}\right]$$

$$= \frac{\sqrt{AB}}{\pi}\left[-\frac{\pi}{2} + A\int_0^{\frac{\pi}{2}} \frac{d\psi}{A - (\alpha\sin^2\psi + \beta\cos^2\psi)} - C\int_0^{\frac{\pi}{2}} \frac{d\psi}{\alpha\sin^2\psi + \beta\cos^2\psi - C}\right]$$

$$= \frac{\sqrt{AB}}{\pi}\left[-\frac{\pi}{2} + \frac{\pi A}{2\sqrt{(A-\alpha)\,(A-\beta)}} - \frac{\pi C}{2\sqrt{(\alpha - C)\,(\beta - C)}}\right]$$

$$= \frac{\sqrt{AB}}{2}\left[-1 + \frac{\pi A}{2\sqrt{A^2 - A\,(A + C - AB) + AC}} - \frac{\pi C}{\sqrt{AC - C\,(A + C - AB) + C^2}}\right]$$

$$= \frac{\sqrt{AB}}{2}\left(-1 + \frac{1}{\sqrt{B}} - \sqrt{\frac{C}{AB}}\right)$$

$$= \tfrac{1}{2}\,(-\sqrt{AB} + \sqrt{A} - \sqrt{C}).$$

$$\therefore\ \mathscr{J}_2 = \frac{1}{2}\left[-J_1\frac{8\pi^2 me}{3FJ_1^2}\psi + J_1 - J_3\right],$$

$$\therefore\ J_1 - J_3 - 2\mathscr{J}_2 = \frac{8\pi^2 me}{3FJ_1}\psi,$$

$$\therefore\ \psi = \frac{3FJ_1}{8\pi^2 me}\,(J_1 - J_3 - 2\mathscr{J}_2)$$

$$= \frac{3F\mathscr{J}_1}{8\pi^2 me}\,(\mathscr{J}_1 - \mathscr{J}_3 - 2\mathscr{J}_2),$$

since $J_1 = \mathscr{J}_1, \quad J_3 = \mathscr{J}_3.$

Therefore the energy in the perturbed orbit is

$$\frac{-2\pi^2 me^4}{J_1^2}+\psi$$

$$=\frac{-2\pi^2 me^4}{\mathcal{J}_1^2}+\frac{3F\mathcal{J}_1}{8\pi^2 me}(\mathcal{J}_1-\mathcal{J}_3-2\mathcal{J}_2).$$

Writing $$\mathcal{J}_1-\mathcal{J}_3-2\mathcal{J}_2=\mathcal{J}_0,$$

the energy $$=\frac{-2\pi^2 me^4}{\mathcal{J}_1^2}+\frac{3F\mathcal{J}_1\mathcal{J}_0}{8\pi^2 me}.$$

The new frequency (previously called ω_F) is

$$\frac{\partial\psi}{\partial\mathcal{J}_0}=\frac{3F\mathcal{J}_1}{8\pi^2 me}, \text{ as before.}$$

The quantum conditions are $\mathcal{J}_1=nh$, $\mathcal{J}_0=kh$, so that the energy of the stationary state h, k is

$$E=\frac{-2\pi^2 me^4}{n^2h^2}+\frac{3Fh^2}{8\pi^2 me}nk, \text{ as before.} \qquad (\S\ 71)$$

This illustrates the application of Bohr's perturbation theory of § 133.

CHAPTER XXI

GENERAL PERTURBATION THEORY

139. *The perturbations of a non-degenerate system*[1]. Suppose the solution of the undisturbed motion found by a separation of variables so that the coordinates are known in terms of action and angle variables J, w. The Hamiltonian function H_0 will be a function of the J's only.

Let the Hamiltonian function H for the disturbed motion be $H_0 + \lambda H_1 + \lambda^2 H_2 + \ldots$, where the parameter λ is of the same order as the ratio of the disturbing forces to those already in action.

First approximation, neglecting λ^2. Substituting the coordinates in H, we have

$$H = H_0(J) + \lambda H_1(J, w),$$

where the H's are used as functional symbols.

In order to find the action and angle variables for the perturbed system, we make the transformation

$$\Sigma (J\delta w + w'\delta J') = \delta S,$$

so that
$$J = \frac{\partial S}{\partial w}, \quad w' = \frac{\partial S}{\partial J'}$$

and S is a function of w, J' (§ 41, equations β).

If Burgers' three conditions (§ 58) are satisfied, viz.

(i) the q's are periodic in w' with period 1,

(ii) H is a function of the J''s only,

(iii) $S - \Sigma w'J'$ is periodic in w' with period 1,

then the J''s are adiabatic invariants for the perturbed system. The quantum conditions will be $J' = n'h$ and the energy H of the stationary states of the perturbed system is then determined.

[1] H. Poincaré, 'Méthodes nouvelles de la Mécanique Celeste,' Paris, 1892–99.

We have then to find the function S by which such a transformation can be effected.

Let $$S = \Sigma wJ' + \lambda S_1(w, J') \quad \ldots\ldots\ldots\ldots\ldots(1),$$

where S_1 is periodic in w with period 1. Then since $w' = \frac{\partial S}{\partial J'}$, it follows that $w' = w +$ (a periodic function of w of period 1). Therefore if w increases by 1, w' increases by 1; and the q's being periodic in the w's of period 1 are also periodic in the w''s of period 1.

Again $(S - \Sigma wJ') - (S - \Sigma w'J')$ has no period in w, w', for it is equal to $\Sigma(w' - w)J'$ and w, w' can only increase by unity together. But from (1) $S - \Sigma wJ' = \lambda S_1(w, J')$ and S_1 is periodic in w with period 1.

Therefore $S - \Sigma wJ'$ is periodic in w with period 1.

Therefore $S - \Sigma w'J'$ is periodic in w (and therefore w') with period 1.

Thus a function S of the form (1) satisfies Burgers' conditions (i) and (iii).

It remains to satisfy (ii), namely, that H shall be a function of the J''s only.

Suppose then that

$$H \equiv H_0(J) + \lambda H_1(J, w) = W(J'), \text{ say.}$$

Then

$$H_0\left(\frac{\partial S}{\partial w}\right) + \lambda H_1\left(\frac{\partial S}{\partial w}, w\right) = W_0(J') + \lambda W_1(J') \ldots \text{ suppose.}$$

Then from (1),

$$H_0\left(J' + \lambda\frac{\partial S_1}{\partial w}\right) + \lambda H_1(J', w) = W_0(J') + \lambda W_1(J'),$$

neglecting λ^2.

Expanding the first term by Taylor's theorem,

$$H_0(J') + \Sigma\lambda\frac{\partial S_1}{\partial w}\left(\frac{\partial H_0}{\partial J}\right)_{J'} + \lambda H_1(J', w) = W_0 + \lambda W_1,$$

where the suffix J' means that J is put equal to J' after differentiation.

$$\therefore \quad W_0 = H_0(J'),$$

and $$W_1 = H_1(J', w) + \Sigma\frac{\partial S_1}{\partial w}\left(\frac{\partial H_0}{\partial J}\right)_{J'} \quad \ldots\ldots\ldots(2).$$

Taking mean values over the unperturbed path (i.e. integrating over a time including a large number of periods of the w's) and denoting the mean value of a function by a bar drawn over it, we have

$$W_1(J') = \overline{H_1(J', w)},$$

since $\overline{\dfrac{\partial S_1}{\partial w}} = 0$ on account of the periodicity of S_1 and therefore of $\dfrac{\partial S_1}{\partial w}$ in the w's.

The right-hand side is a function of the J''s only, so that W_1 is found.

Thus to a first approximation, the energy of the disturbed path is greater than that of the undisturbed path by the mean value of the 'perturbing function,' λH_1, taken over the undisturbed path (cf. § 133, equation 5).

Thus $$H = W_0(J') + \lambda W_1(J'),$$

and the energy of the stationary states of the perturbed system is

$$W_0(n'h) + \lambda W_1(n'h),$$

where $$J_\tau' = n_\tau' h.$$

140. *Nature of the perturbations.* From (2),

$$H_1(J', w) - W_1(J') = -\Sigma \frac{\partial S_1}{\partial w}\left(\frac{\partial H_0}{\partial J}\right)_{J'}$$
$$= \Sigma A_\tau e^{2\pi i(\tau w)},$$

since the right-hand side is periodic in the w's with period 1. Here (τw) denotes $\tau_1 w_1 + \ldots + \tau_s w_s$, where s is the number of coordinates or frequencies (the system being non-degenerate); also A_τ is a function of J'. Also S_1 can be expanded as $\Sigma B_\tau e^{2\pi i(\tau w)}$.

But $$\Sigma \frac{\partial S_1}{\partial w}\left(\frac{\partial H_0}{\partial J}\right)_{J'} = \Sigma \frac{\partial S_1}{\partial w}.\nu,$$

where $$\nu = \left(\frac{\partial H_0}{\partial J}\right)_{J'}$$

is an undisturbed frequency, with J' written for J.

$$\therefore\ \Sigma \frac{\partial S_1}{\partial w}\left(\frac{\partial H_0}{\partial J}\right)_{J'} = \Sigma 2\pi i\,(\tau\nu)\,B_\tau\,e^{2\pi i(\tau w)},$$

$$\therefore\ A_\tau = -\,2\pi i\,(\tau\nu)\,B_\tau,$$

$$\therefore\ S_1 = -\,\Sigma' \frac{1}{2\pi i}\frac{A_\tau}{(\tau\nu)}\,e^{2\pi i(\tau w)},$$

all the terms being finite since $(\tau\nu)$ is not equal to zero for any combination of the τ's (as it would be for some combination, if the system were degenerate).

$$\therefore\ w' = \frac{\partial S}{\partial J'} = w + \lambda \frac{\partial S_1}{\partial J'} = w - \frac{\lambda}{2\pi i}\Sigma' \frac{\partial}{\partial J'}\left\{\frac{A_\tau}{(\tau\nu)}\right\} e^{2\pi i(\tau w)}.$$

Thus the w''s oscillate about the values w, with amplitude of order λ. The equation

$$J = \frac{\partial S}{\partial w} = J' + \lambda \frac{\partial S_1}{\partial w}$$

leads to the same conclusion about J'.

Thus in a non-degenerate case there are no 'secular perturbations'; these, as we have seen in the case of the Keplerian orbit, are associated with degenerate systems.

141. *Second approximation*[1], *including* λ^2. As before

$$H = H_0 + \lambda H_1 + \lambda^2 H_2 + \dots,$$

and $$S = \Sigma w J' + \lambda S_1(J', w) + \lambda^2 S_2(J', w)\dots.$$

$$\therefore\ H_0\left(\frac{\partial S}{\partial w}\right) + \lambda H_1\left(\frac{\partial S}{\partial w}, w\right) + \lambda^2 H_2\left(\frac{\partial S}{\partial w}, w\right)$$
$$= W(J')$$
$$= W_0(J') + \lambda W_1(J') + \lambda^2 W_2(J').$$

$$\therefore\ H_0\left(J' + \lambda\frac{\partial S_1}{\partial w} + \lambda^2\frac{\partial S_2}{\partial w}\right) + \lambda H_1\left(J' + \lambda\frac{\partial S_1}{\partial w}, w\right) + \lambda^2 H_2(J', w)$$
$$= W_0 + \lambda W_1 + \lambda^2 W_2, \text{ up to } \lambda^2.$$

Therefore expanding H_0, H_1, we have

$$H_0(J') + \Sigma\left(\lambda\frac{\partial S_1}{\partial w} + \lambda^2\frac{\partial S_2}{\partial w}\right)\left(\frac{\partial H_0}{\partial J}\right)_{J'} + \frac{\lambda^2}{2!}\Sigma\frac{\partial S_1}{\partial w_r}\frac{\partial S_1}{\partial w_s}\left(\frac{\partial^2 H_0}{\partial J_r \partial J_s}\right)_{J'}$$
$$+ \lambda\left[H_1(J', w) + \lambda\frac{\partial S_1}{\partial w}\left(\frac{\partial H_1}{\partial J}\right)_{J'}\right] + \lambda^2 H_2(J', w)$$
$$= W_0 + \lambda W_1 + \lambda^2 W_2.$$

[1] M. Born and W. Pauli, Jr, Zeitschr. für Physik, **10**, p. 137, 1922.

The *new* equation, arising from the λ^2, is

$$\Sigma \frac{\partial S_2}{\partial w}\left(\frac{\partial H_0}{\partial J}\right)_{J'} + \tfrac{1}{2}\Sigma \frac{\partial S_1}{\partial w_r}\frac{\partial S_1}{\partial w_s}\left(\frac{\partial^2 H_0}{\partial J_r \partial J_s}\right)_{J'} + \frac{\partial S_1}{\partial w}\left(\frac{\partial H_1}{\partial J}\right)_{J'} + H_2(J', w) = W_2(J').$$

The first two terms are periodic in the w's, because $\frac{\partial S_1}{\partial w}$, $\frac{\partial S_2}{\partial w}$ are, and the coefficients $\left(\frac{\partial H_0}{\partial J}\right)_{J'}$, $\left(\frac{\partial^2 H_0}{\partial J_r \partial J_s}\right)_{J'}$ are functions of J' only (and so independent of w).

The third term is not periodic, because $\left(\frac{\partial H_1}{\partial J}\right)_{J'}$ is

$$\left\{\frac{\partial}{\partial J} H_1(J, w)\right\}_{J'}$$

and contains w.

Therefore taking mean values of both sides, as before,

$$\Sigma \overline{\frac{\partial S_1}{\partial w}\left(\frac{\partial H_1}{\partial J}\right)_{J'}} + \overline{H_2(J', w)} = W_2(J').$$

Thus the energy is

$$W(J') = W_0(J') + W_1(J') + W_2(J'),$$

where

$$W_0(J') = H_0(J'),$$
$$W_1(J') = \overline{H_1(J', w)},$$
$$W_2(J') = \overline{H_2(J', w)} + \Sigma \overline{\left(\frac{\partial H_1}{\partial J}\right)_{J'} \frac{\partial S_1}{\partial w}}.$$

The quantum conditions as before, $J' = n'h$ determine the stationary states, to order λ^2.

142. *Application to the anharmonic oscillator*. The Hamiltonian function is

$$H = c_0 + \frac{p_1^2}{2m} + \tfrac{1}{2}m(2\pi\omega)^2 x^2 + cx^3 + bx^4 + \dots \text{ (cf. p. 169, line 7)}$$

$$= H_0 + \lambda H_1 + \lambda^2 H_2 \dots,$$

where

$$\left.\begin{aligned} H_0 &= c_0 + \frac{p_1^2}{2m} + \tfrac{1}{2}m(2\pi\omega)^2 x^2 \\ \lambda H_1 &= cx^3 \\ \lambda^2 H_2 &= bx^4 \end{aligned}\right\}.$$

The motion is the perturbation of the motion given by H_0, the perturbing function being $\lambda H_1 + \lambda^2 H_2$.

The *undisturbed* motion is given by

$$H_0 = c_0 + \frac{p_1^2}{2m} + \tfrac{1}{2}m(2\pi\omega)^2 x^2 = \alpha_1,$$

$$J = \oint p_1 dx = \oint \sqrt{2m(\alpha_1 - c_0) - m^2(2\pi\omega)^2 x^2}\, dx$$

$$= \oint 2\pi m\omega \sqrt{(k^2 - x^2)}\, dx,$$

where
$$k^2 = \frac{2(\alpha_1 - c_0)}{m(2\pi\omega)^2},$$

$$\therefore J = 2\pi m\omega . 2\int_{-k}^{k} \sqrt{k^2 - x^2}\, dx.$$

Writing
$$x = \sin\theta,$$

$$J = 4\pi m\omega \int_{-\frac{\pi}{2}}^{\frac{\pi}{2}} k^2 \cos^2\theta d\theta$$

$$= 2\pi^2 m\omega k^2$$

$$= \frac{\alpha_1 - c_0}{\omega}, \text{ writing in } k^2,$$

$$\therefore H_0 = \alpha_1 = c_0 + \omega J.$$

Therefore the frequency $\frac{\partial H_0}{\partial J}$ is equal to ω.

Also $S = \int \frac{\partial S}{\partial x} dx = \int p_1 dx = \int \sqrt{2m(\alpha_1 - c_0) - m^2(2\pi\omega)^2 x^2}$

$$= \int \sqrt{2m\omega J - m^2(2\pi\omega)^2 x^2} . dx,$$

and
$$w = \frac{\partial S}{\partial J} = m\omega \int \frac{dx}{\sqrt{2m\omega J - m^2(2\pi\omega)^2 x^2}}$$

$$= \frac{1}{2\pi} \sin^{-1}\left\{x\sqrt{\frac{2\pi^2 m\omega}{J}}\right\},$$

$$\therefore x = \left(\frac{J}{2\pi^2 m\omega}\right)^{\frac{1}{2}} \sin 2\pi w,$$

and $p_1 = \sqrt{2m\omega J - m^2(2\pi\omega)^2 x^2} = (2m\omega J)^{\frac{1}{2}} \cos 2\pi w.$

Thus the undisturbed motion is simple harmonic of frequency ω.

Using the value of x, we have for the *disturbed* motion,

$$H = \underbrace{c_0 + \omega J}_{H_0} + \underbrace{c\left(\frac{J}{2\pi^2 m\omega}\right)^{\frac{3}{2}} \sin^3 2\pi w}_{\lambda H_1} + \underbrace{b\left(\frac{J}{2\pi^2 m\omega}\right)^2 \sin^4 2\pi w}_{\lambda^2 H_2} + \ldots.$$

From the general theory

$$H = W_0(J') + \lambda W_1(J') + \lambda^2 W_2(J'),$$

where
$$W_0(J') = H_0(J') = c_0 + \omega J',$$

$$\lambda W_1(J') = \overline{\lambda . H_1(J', w)} = c\left(\frac{J'}{2\pi^2 m\omega}\right)^{\frac{3}{2}} \overline{\sin^3 2\pi w}$$

$$= c\left(\frac{J'}{2\pi^2 m\omega}\right)^{\frac{3}{2}} \left(\frac{3\,\overline{\sin 2\pi w} - \overline{\sin 6\pi w}}{8}\right) = 0.$$

From the general condition (2) of § 139,

$$W_1(J') = H_1(J', w) + \Sigma \frac{\partial S_1}{\partial w}\left(\frac{\partial H_0}{\partial J}\right)_{J'}.$$

Here this becomes, since W_1 has just been proved 0,

$$0 = c\left(\frac{J'}{2\pi^2 m\omega}\right)^{\frac{3}{2}} \sin^3 2\pi w + \lambda \frac{\partial S_1}{\partial w}\,\omega,$$

$$\therefore \frac{\partial S_1}{\partial w} = -\frac{c}{\lambda\omega}\left(\frac{J'}{2\pi^2 m\omega}\right)^{\frac{3}{2}} \sin^3 2\pi w.$$

From the general equation

$$W_2(J') = \overline{H_2(J', w)} + \overline{\Sigma\left(\frac{\partial H_1}{\partial J}\right)_{J'} \frac{\partial S_1}{\partial w}},$$

we have here

$$W_2(J') = \frac{b}{\lambda^2}\left(\frac{J'}{2\pi^2 m\omega}\right)^2 \overline{\sin^4 2\pi w}$$

$$+ \frac{c}{\lambda}\frac{3}{2}\frac{J'^{\frac{1}{2}}}{(2\pi^2 m\omega)^{\frac{3}{2}}} \overline{\sin^3 2\pi w \left[-\frac{c}{\lambda\omega}\left(\frac{J'}{2\pi^2 m\omega}\right)^{\frac{3}{2}} \sin^3 2\pi w\right]},$$

$$\therefore \lambda^2 W_2(J') = b\left(\frac{J'}{2\pi^2 m\omega}\right)^2 \overline{\sin^4 2\pi w} - \frac{3c^2}{2\omega}\frac{J'^2}{(2\pi^2 m\omega)^3} \overline{\sin^6 2\pi w},$$

and since $\overline{\sin^4 2\pi w} = \frac{3}{8}$, and $\overline{\sin^6 2\pi w} = \frac{5}{16}$,

$$\therefore\ \lambda^2 W_2(J') = \frac{3b}{8}\left(\frac{J'}{2\pi^2 m\omega}\right)^2 - \frac{15c^2}{32\omega}\frac{J'^2}{(2\pi^2 m\omega)^3}.$$

Therefore for the perturbed motion (the anharmonic oscillator)

$$H = W_0 + \lambda W_1 + \lambda^2 W_2$$

$$= c_0 + \omega J' - \frac{15c^2}{32\omega}\frac{J'^2}{(2\pi^2 m\omega)^3} + \frac{3b}{8}\left(\frac{J'}{2\pi^2 m\omega}\right)^2.$$

This with the omission of the term in b (omitted in § 118) is the result (5) found in § 120.

143. *The perturbations of a degenerate system.* For such a system having s coordinates, there is a smaller number u of independent frequencies. The frequencies are $\omega_1 \ldots \omega_u$ and the other ω's are zero. Thus the angle variables are

$$w_1 = \omega_1 t + \delta_1, \ldots, \quad w_u = \omega_u t + \delta_u, \quad w_{u+1} = \delta_{u+1}, \ldots, w_s = \delta_s.$$

Let w_α denote a typical w ($n = 1$ to u),

w_β „ „ w ($n = u + 1$ to s),

and let J_α, J_β correspond. The w_β's are constants and the w_α's vary uniformly with the time.

To a first approximation $H = H_0 + \lambda H_1$, and on substituting the coordinates in H we find

$$H = H_0(J_\alpha) + \lambda H_1(J_\alpha, J_\beta, w_\alpha, w_\beta).$$

By a contact transformation

$$\Sigma\,(J_\alpha dw_\alpha - w_\alpha' dJ_\alpha') = dS,$$

we can as before remove the w_α's from H_1, and obtain

$$H = W_0(J_\alpha') + \lambda W_1(J_\alpha', J_\beta, w_\beta).$$

Then since $\dot{w} = \dfrac{\partial H}{\partial J}$ and $-\dot{J} = \dfrac{\partial H}{\partial w}$,

$$\dot{w}_\beta = \frac{\partial}{\partial J_\beta}\left[W_0(J_\alpha') + \lambda W_1(J_\alpha', J_\beta, w_\beta)\right]$$

$$= \lambda\frac{\partial}{\partial J_\beta} W_1(J_\alpha', J_\beta, w_\beta),$$

and

$$\dot{J}_\beta = -\lambda\frac{\partial}{\partial w_\beta} W_1(J_\alpha', J_\beta, w_\beta).$$

These are of the Hamiltonian form for $(s-u)$ degrees of freedom, with $\lambda W_1(J_\alpha, 'J_\beta, w_\beta)$ as the Hamiltonian function; and they describe the 'secular perturbations' which arise from the degenerate character of the original system.

Introducing new action variables given by $\mathscr{J} = \oint J_\beta dw_\beta$, this new Hamiltonian function $\lambda W_1(J_\alpha', J_\beta, w_\beta)$ can be expressed in terms of J_α', $\mathscr{J}$ only; denote this function by $\psi(J_\alpha', \mathscr{J})$.

This introduces $(s-u)$ $\mathscr{J}$'s, and if the number of independent frequencies $\left(\text{given by } \frac{\partial\psi}{\partial\mathscr{J}}\right)$ of the secular perturbations is u', then by a linear transformation we can express ψ in terms only of new $\mathscr{J}'$'s which are u' in number.

Finally, H is expressed in terms of J_α' (u in number), $\mathscr{J}'$ (u' in number); then $J_\alpha' = nh$, $\mathscr{J}' = n'h$ determine the stationary states, which have the u frequencies of the unperturbed orbit together with the u' new frequencies of the secular perturbations. [For instance in the case of a Keplerian orbit disturbed by an electric field, there were three coordinates and one frequency in the undisturbed orbit. Thus there were one w_α and two w_β's and $u = 1$. In the perturbations there would be two $\mathscr{J}$'s and two equal frequencies and therefore only one $\mathscr{J}'$; so that $u' = 1$.]

The theory owes much to the work of Bohr[1], Kramers[2] and Epstein[3] in the past few years. Born[4] and his collaborators have extended it to higher orders of approximation. A formal statement of the present state of the theory has been given by Bohr[5] in an essay on 'The fundamental postulates of the quantum theory.'

[1] N. Bohr, 'The Quantum Theory of Line Spectra,' Parts I, II, Copenhagen, 1918.

[2] H. A. Kramers, 'Intensities of Spectral Lines,' Copenhagen, 1919; Zeitschr. für Physik, **3**, p. 199, 1920; **13**, p. 312, 1923.

[3] P. S. Epstein, Zeitschr. für Physik, **8**, pp. 211, 305; **9**, p. 92, 1922.

[4] M. Born and E. Brody, Zeitschr. für Physik, **6**, p. 140, 1921; M. Born and W. Pauli, Jr, Zeitschr. für Physik, **10**, p. 137, 1922; M. Born and W. Heisenberg, Zeitschr. für Physik, **14**, p. 44, and **16**, p. 229, 1923; Ann. d. Physik, **74**, p. 1, 1924.

[5] N. Bohr, Proc. Camb. Phil. Soc. Supplement, 1924.

CHAPTER XXII

THE EFFECT OF ELECTRIC AND MAGNETIC FIELDS ON THE SPECTRA OF ELEMENTS OF HIGHER ATOMIC NUMBER. RECENT DEVELOPMENTS OF THE QUANTUM THEORY: THE DOUBLET THEORIES OF SOMMERFELD AND LANDÉ; THE DISPERSION THEORY OF KRAMERS; THE QUANTUM-KINEMATICS OF HEISENBERG

144. *The effect of an electric field on series spectra.* The lines of the series spectra are in their general aspect accounted for by supposing the outer electron to revolve in a central field due to the core, the frequency in the osculating orbit being ω_n and the frequency of revolution of the orbit itself ω_k. If for the main purpose we neglect the multiple structure, then Bohr's theory leads to the formula $Rhc/[n + \psi(k)]^2$ for the negative energy of a series 'term,' where from the mode of derivation of the formula (§ 90), $\psi(k)$ increases with ω_k. But $\psi(k)$ measures the deviation of the term from the hydrogen term Rhc/n^2, so that this deviation is a measure of the rapidity of the changes in the position of the orbit.

It has been seen (§ 64) that the relativity correction for the H orbit leads to an orbit of this kind, with a very small value of ω_k; and that this correction is equivalent to the effect of a *small* central field (§ 66). Thus it would seem that the dynamical problem of the effect of a strong electric field upon the spectrum of an element of higher atomic number (where ω_k is not small) would be the same as that of the effect of a *weak* electric field on the H atom allowing for relativity, the problems differing not in kind but only in degree.

This problem of the effect of a *weak* electric field (F) on the fine structure of the hydrogen lines was solved by

Kramers[1], who found that no resolution of the lines into components whose displacement is proportional to F occurs (as is the case with a *strong* field (§ 74)). But new lines of frequencies $\omega_n \pm \omega_k$ appear, corresponding to 'beats' in acoustics, because transitions not possible for the unperturbed orbit become possible for the perturbed one. The Fourier series for the coordinates in the perturbed orbit contain new terms which mean that new transitions for which $\Delta k = 0$ or 2 are now possible, and these give rise to the new lines.

This theory was in accord with the observations of Stark[2] on the series lines of He and Li, who found that the ordinary Stark effect of the hydrogen type was not measurable at all, but that new lines with frequencies equal to the sum or difference of the old frequencies appeared.

In a second paper, already referred to (§ 133), Kramers[3], using the Delaunay elements in the manner of (§ 138), solved the problem of the effect of *any* electric field (F) on the fine structure of the hydrogen lines. He was thus able to trace how the effect for a weak field gradually changes to the normal Stark effect as the field F increases.

But it must be recognised that these results are only a first step in the theory of the influence of electric fields upon series spectra; the comparison between the perturbing effect of the field due to the core and the action of a central field (even with an axis of symmetry, to allow for the doublets (§ 87)) is only an imperfect analogy. The general problem requires a closer consideration of the disturbance of the core itself during the perturbation of the outer electron.

145. *The effect of a magnetic field on series spectra (the anomalous Zeeman effect).* The theory of the Zeeman effect given in § 79 is of so general a character that it might be expected to hold for spectra other than that of hydrogen; the motion in a stationary state with the magnetic field should

[1] H. A. Kramers, 'Intensities of Spectral Lines,' p. 26, Copenhagen, 1919.

[2] J. Stark, Ann. d. Phys. **48**, p. 210, 1915.

[3] H. A. Kramers, Zeitschr. für Physik, **3**, p. 199, 1920.

on this theory be that in a stationary state without the field, together with the Larmor precession ω_H; each line should be resolved into a triplet (or doublet).

It is well known that this is not the case for the series spectra, though in the case of He and Li, where the spectra consist of single lines or very narrow doublets, there is to a high approximation no departure from the normal effect.

But for the alkalis of higher atomic number, where the lines consist of doublets of very considerable width (§ 86), the effects are far more complex. Each member of such a doublet is resolved into a large number of components whose displacements are proportional to the magnetic field (H), but different for the two members of the doublet (anomalous Zeeman effect)[1]. As H increases and the displacements of the components become of the same order of magnitude as the original width of the doublet, the resolutions undergo gradual changes, until for large fields the components of both members flow together into the normal Zeeman triplet (Paschen-Back effect)[2].

It is apparent that Larmor's theorem does not hold for these changes, but that theorem depends for its validity upon the field due to the core having the same specification with respect to the two sets of axes concerned (see the end of § 77); it is hardly to be expected that such a condition would be satisfied all the time by a complex system, such as a core, consisting of a nucleus and many electrons.

146. *Sommerfeld's theory of the X-ray doublets*. Sommerfeld[3] accounted for the K, L doublets of the X-ray spectra (cf. § 105) by supposing that the L orbits have an energy difference due to relativity given by the same type of formula as that for the fine structure of hydrogen. The L orbits are

[1] For full details of the 'anomalous' Zeeman effect see C. W. Voigt, Magneto- und Elektro-Optik, Leipzig, 1908 (classical theory); E. Back u. A. Landé, 'Zeemaneffekt und Multiplettstruktur der Spektrallinien,' 1924; A. Sommerfeld, 'Atomic Structure and Spectral Lines' (Engl. Tr.), chaps. VI, VII, 1923; N. Bohr, Proc. Phys. Soc. London, **35**, p. 275, 1923.

[2] F. Paschen u. E. Back, Ann. d. Phys. **39**, p. 897, 1912.

[3] A. Sommerfeld, 'Atomic Structure and Spectral Lines,' p. 501, 1923.

effectively described under the action of the nuclear charge Ze whose effect is modified to $Z_i e$ by the screening effect of the core electrons [$Z_i = Z - 1$ for the K lines (§ 99) and $Z_i = Z - s$ for the L lines where s is about 4].

From § 62, the energy of an n_k orbit for hydrogen is given by

$$W = Rhc\left[\frac{1}{n^2} + \frac{\mu^2}{n^4}\left(\frac{n}{k} - \frac{3}{4}\right)\right],$$

where $$\mu = 2\pi e^2/ch.$$

Here the effective nuclear charge is $Z_i e$ and the 'effective' quantum number (§ 85) is n_i, so that

$$W = RhcZ_i^2\left[\frac{1}{n_i^2} + \frac{\mu^2 Z_i^2}{n_i^4}\left(\frac{n_i}{k} - \frac{3}{4}\right)\right].$$

The doublet separation is due to the difference of the W's for the two n_i orbits for which the second quantum numbers are $k, k-1$.

$$\therefore\ \Delta W = \frac{Rhc\mu^2 Z_i^4}{n_i^4}\left(\frac{n_i}{k-1} - \frac{n_i}{k}\right)$$

$$= \frac{Rhc\mu^2 Z_i^4}{k(k-1)\,n_i^3}.$$

The doublet separation

$$\Delta\nu_i = \frac{1}{h}\Delta W = \frac{Rc\mu^2 Z_i^4}{k(k-1)\,n_i^3}.$$

For the L doublet n_i is about 2 and k is also, so that

$$\Delta\nu_i = \frac{Rc\mu^2 Z_i^4}{2^4}\ \text{approx.}$$

This formula enables Z_i to be calculated from observations of $\Delta\nu_i$ and was found by Sommerfeld to hold quite well for the elements from Nb (41) to U (92). The value of Z_i was found to be $(Z - 3{\cdot}5)$ approx.

147. *Landé's theory of the optical doublets.* Landé[1] supposes that the time t_i spent in describing the inner orbital loop

[1] A. LANDÉ, Zeitschrift für Phys. **25**, p. 46, 1924.

(shown in the figure on p. 121) within the core is small compared with the time t_a spent in describing the outer orbital loop. The latter is approximately a Keplerian ellipse described under an effective nuclear charge $Z_a e$, where $Z_a = 1$ for a neutral atom A, $Z_a = 2$ for A_+, $Z_a = 3$ for A_{++}, and so on. Effectively t_a is the time of revolution in the orbit supposed completely Keplerian.

The inner orbital loop is described under the control of the nuclear charge modified just as in Sommerfeld's theory of the X-ray doublet.

Landé supposes that $\Delta\nu$ for the optical lines is found approximately by multiplying the $\Delta\nu_i$ of Sommerfeld by the fraction of the whole period that the electron spends in the inner loop (where it is under the action which gave rise to $\Delta\nu_i$) or

$$\Delta\nu = \frac{t_i}{t_a}\Delta\nu_i.$$

Also the period t in the orbit is

$$\frac{n^3 h^3}{4\pi^2 m e^4 Z^2} \qquad (\S\ 30),$$

so that

$$\frac{t_i}{t_a} = \frac{n_i^3}{Z_i^2}\cdot\frac{Z_a^2}{n_a^3},$$

where n_a is the 'effective' quantum number of § 85 and is the $n - a_k$ of Rydberg's formula.

Therefore $\Delta\nu$ for the optical lines

$$= \Delta\nu_i \frac{Z_a^2}{Z_i^2}\cdot\frac{n_i^3}{n_a^3}$$

$$= \frac{Rc\mu^2 Z_i^4}{k(k-1)\,n_i^3}\cdot\frac{Z_a^2}{Z_i^2}\cdot\frac{n_i^3}{n_a^3}$$

$$= \frac{Rc\mu^2 Z_i^2 Z_a^2}{k(k-1)\,n_a^3}.$$

This is Landé's formula for the optical doublets, giving a quantitative result for the dependence of the separation $\Delta\nu$

upon atomic number (cf. § 86, where the separations are given for the alkali metals).

148. *The dispersion theory of Kramers.* It is well known that the phenomena of dispersion and absorption of monochromatic light by a gas can be accounted for by the assumption that an atom, when exposed to radiation, becomes a source of secondary spherical waves, which are coherent with the incident waves.

The classical theory of electrodynamics, in which the atom is assumed to contain electrons elastically bound to positions of equilibrium, leads to a dispersion theory which agrees with the results of experiment, not only for normal dispersion, but also in the case of anomalous dispersion near the absorption lines. Yet the search for an exact explanation of dispersion phenomena on the classical theory encounters difficulties bound up with the difficulties of the explanation of the spectrum itself which were insuperable before the advent of the quantum theory.

Kramers therefore attacked the problem of the scattering and dispersion of incident light by an atom as pictured on the quantum theory, that is one whose spectrum is due not to elastically vibrating electrons, but to transitions from one stationary state to another.

The correspondence principle requires that a quantum theory formula of dispersion should asymptotically tend to the classical one for large quantum numbers; it was by its use, and by certain new concepts given in a joint paper by Bohr, Kramers, and Slater[1], that the new dispersion formula was derived.

One of these concepts is that the atom in a given stationary state reacts to external influences in the same way as a virtual radiation field whose frequencies are those of the various possible transitions to other stationary states given by Bohr's frequency condition.

[1] N. Bohr, H. A. Kramers and J. C. Slater, Phil. Mag. **47**, p. 785, **1924**.

149. On the classical theory, an oscillating bipole whose electric moment is the real part of $\mathbf{C}e^{2\pi i\nu t}$ radiates energy per second of amount

$$\frac{(2\pi\nu)^4}{3c^3}(\mathbf{C}\bar{\mathbf{C}}) \quad \text{.......................(1)},$$

where $\bar{\mathbf{C}}$ is the vector conjugate to $\mathbf{C}$.

Kramers, using the concept referred to above, assumed that the atom in a given stationary state behaves like an oscillating bipole whose moment is

$$\Sigma_q \mathbf{A}_q e^{2\pi i\nu_q t} \quad \text{.......................(2)},$$

where the ν_q's are the frequencies corresponding to all possible transitions to other stationary states, and the amplitude vector $\mathbf{A}_q$ is connected, on account of (1), with the Einstein probability coefficient a_q by the relation

$$a_q(h\nu_q) = \frac{(2\pi\nu_q)^4}{3c^3}(\mathbf{A}_q\bar{\mathbf{A}}_q) \text{.................(3)}.$$

[a_q is the number of transitions per second of the type corresponding to the frequency ν_q (cf. § 26), so that $a_q(h\nu_q)$ is the emission of energy per second.]

The formulae (2), (3) may be looked upon as representing the atom through its spectrum, ν_q being the frequency of a spectral line and a_q representing the intensity of the line, and these are both quantities, associated with the atom, which can be observed experimentally.

150. Kramers and Heisenberg[1] then proceeded to consider a multiply periodic system which would cause the spectrum defined by the formulae (2), (3).

The frequencies (ω) and amplitude vectors ($\mathbf{C}$) of this system are different from the frequencies (ν_q) and intensity vectors ($\mathbf{A}_q$) of the spectrum, but for large quantum numbers $\omega \to \nu_q$ and $\mathbf{C} \to \mathbf{A}_q$ asymptotically, by the correspondence principle.

Let the multiply periodic system which is the origin of the

[1] H. A. Kramers and W. Heisenberg, Zeitschr. für Phys. **31**, p. 681, 1925.

spectrum be defined by the action and angle variables $I_1, \ldots I_s$, $w_1, \ldots w_s$, and let its electric moment be

$$\Sigma_\tau \tfrac{1}{2}\mathbf{C}_{\tau_1 \ldots \tau_s} e^{2\pi i(\tau_1 w_1 + \ldots + \tau_s w_s)},$$

where the $\mathbf{C}$'s depend upon the I's only.

The frequencies of the system are of the type

$$\begin{aligned}\omega &\equiv \tau_1\omega_1 + \ldots + \tau_s\omega_s \\ &= \tau_1\frac{\partial H}{\partial I_1} + \ldots + \tau_s\frac{\partial H}{\partial I_s} \\ &= \frac{\partial H}{\partial I}, \text{ suppose,}\end{aligned}$$

where
$$\frac{\partial}{\partial I} \equiv \tau_1\frac{\partial}{\partial I_1} + \ldots + \tau_s\frac{\partial}{\partial I_s}.$$

The problem now is to consider the effect of incident monochromatic light represented by the electric vector $\mathbf{E}e^{2\pi i\nu t}$ upon this system, which is effectively the atom itself.

By the use of an infinitesimal contact transformation (§ 48) to new variables J, v, the perturbing effect of the incident light was calculated, and the *change* of the electric moment of the system found to be

$$\Sigma_\tau \Sigma_{\tau'} \frac{1}{4}\left[\frac{\partial \mathbf{C}}{\partial J'} \cdot \frac{(\mathbf{EC}')}{\omega' + \nu} - \mathbf{C}\frac{\partial}{\partial J}\left\{\frac{(\mathbf{EC}')}{\omega' + \nu}\right\}\right] e^{2\pi i(\omega + \omega' + \nu)t} \quad (4),$$

where
$$\begin{aligned}\frac{\partial}{\partial J} &= \tau_1\frac{\partial}{\partial J_1} + \ldots + \tau_s\frac{\partial}{\partial J_s}, \\ \frac{\partial}{\partial J'} &= \tau_1{}'\frac{\partial}{\partial J_1} + \ldots + \tau_s{}'\frac{\partial}{\partial J_s}, \\ \omega &= \tau_1\omega_1 + \ldots + \tau_s\omega_s, \\ \omega' &= \tau_1{}'\omega_1 + \ldots + \tau_s{}'\omega_s,\end{aligned}$$

and $\mathbf{C}'$ is $\mathbf{C}$ in which τ' is written for τ.

Thus the radiation from the atom due to the incidence of the light of frequency ν consists of waves of frequencies $\nu + \omega + \omega'$, where ω, ω' are any two periods (positive or negative) of the undisturbed system.

The term for which $\omega + \omega' = 0$, reduces to the form

$$\Sigma_\tau \frac{1}{4}\frac{\partial}{\partial J}\left[\frac{\mathbf{C}\,(\mathbf{E}\bar{\mathbf{C}})}{\omega - \nu} + \frac{\bar{\mathbf{C}}\,(\mathbf{E}\mathbf{C})}{\omega + \nu}\right] e^{2\pi i \nu t} \quad \ldots\ldots\ldots\ldots(5),$$

because $\omega + \omega' = 0$ requires $\tau' = -\tau$ and the double summation reduces to a single one; also

$$\mathbf{C}' = \mathbf{C}_{\tau_1'\tau_2'\ldots} = \mathbf{C}_{-\tau_1, -\tau_2, \ldots} = \bar{\mathbf{C}},$$

where $\bar{\mathbf{C}}$ is conjugate to $\mathbf{C}$; and $\frac{\partial}{\partial J'} = -\frac{\partial}{\partial J}$.

This term has thus the same frequency as the incident light, and is also coherent (i.e. in the same phase) with it, because the vectors $\mathbf{C}$, $\bar{\mathbf{C}}$ are multiplied together, and the term is $\mathbf{E}e^{2\pi i \nu t}$ multiplied by a real factor.

This term therefore represents the dispersion and absorption.

151. This classical formula (5) has now to be generalised into a quantum theory formula expressed in terms of the $\mathbf{A}_q$'s and the ν_q's, which will asymptotically tend to the formula (5) for large quantum numbers.

The frequencies of the periodic system which represents the working of the atom are $\omega = \frac{\partial H}{\partial J}$, and of the spectrum characteristic of the atom are $\nu_q = \frac{\Delta H}{h}$, where ΔH is the change of H due to a transition in which $J_1, J_2, \ldots$ change by $\tau_1 h, \tau_2 h, \ldots$ and the corresponding quantum numbers change by $\tau_1, \tau_2, \ldots$ (since $J = nh$).

$$\begin{aligned} \therefore\ \nu_q = \frac{\Delta H}{h} &= \frac{\Delta H}{\Delta J_1}\frac{\Delta J_1}{h} + \ldots \\ &= \frac{\Delta H}{\Delta J_1}\cdot\frac{h\Delta n_1}{h} + \ldots \\ &= \frac{\Delta H}{\Delta J_1}\cdot\tau_1 + \frac{\Delta H}{\Delta J_2}\tau_2 + \ldots, \end{aligned}$$

while

$$\omega = \frac{\partial H}{\partial J_1}\cdot\tau_1 + \ldots.$$

For large quantum numbers the difference quotients $\frac{\Delta H}{\Delta J_1}, \ldots$ become differential coefficients $\frac{\partial H}{\partial J_1}, \ldots$, so that $\nu_q \to \omega$.

When the quantum numbers are not large, ν_q is derived from ω by applying the operator $\frac{\Delta}{h}$ instead of the operator $\frac{\partial}{\partial J}$ to H. So it is presumed that $\mathbf{A}_q$ is to be derived from $\mathbf{C}$, when the quantum numbers are not large, by the same process.

Replacing $\frac{\partial}{\partial J}$ in the formula (5) by the operator $\frac{\Delta}{h}$, where Δ is the effect of changing each J by τh, a quantum theory formula is deduced, which is

$$\sum_a \frac{1}{4h}\left\{\frac{\mathbf{A}_a(\mathbf{E}\bar{\mathbf{A}}_a)}{\nu_a-\nu}+\frac{\bar{\mathbf{A}}_a(\mathbf{E}\mathbf{A}_a)}{\nu_a+\nu}\right\}e^{2\pi i\nu t}$$
$$-\sum_e \frac{1}{4h}\left\{\frac{\mathbf{A}_e(\mathbf{E}\bar{\mathbf{A}}_e)}{\nu_e-\nu}+\frac{\bar{\mathbf{A}}_e(\mathbf{E}\mathbf{A}_e)}{\nu_e+\nu}\right\}e^{2\pi i\nu t} \quad \ldots\ldots(6),$$

where the first sum is for all the frequencies ν_a for which energy is absorbed in the transition, and the second for all frequencies ν_e for which energy is emitted in the transition, from the given stationary state of the atom.

152. In the special case where the vector $\mathbf{E}$ and all the vectors $\mathbf{A}_a$, $\mathbf{A}_e$ are real and parallel to one another, i.e. where the incident light is linearly polarised and where the electric vector in all the transitions is parallel to the light vector, formula (6) becomes

$$\sum_a \frac{1}{2h}\left\{\frac{\mathbf{E}\mathbf{A}_a\bar{\mathbf{A}}_a\nu_a}{\nu_a^2-\nu^2}\right\}e^{2\pi i\nu t}-\sum_e \frac{1}{2h}\left\{\frac{\mathbf{E}\mathbf{A}_e\bar{\mathbf{A}}_e}{\nu_e^2-\nu^2}\right\}e^{2\pi i\nu t};$$

and introducing the Einstein coefficients a by the use of equation (3), viz.

$$a_q(h\nu_q)=\frac{(2\pi\nu_q)^4}{3c^3}\mathbf{A}_q\bar{\mathbf{A}}_q,$$

it reduces to

$$\left[\sum_a \frac{\mathbf{E}a_a 3c^3}{32\pi^4\nu_a^2(\nu_a^2-\nu^2)}-\sum_e \frac{\mathbf{E}a_e 3c^3}{32\pi^4\nu_e^2(\nu_e^2-\nu^2)}\right]e^{2\pi i\nu t}.$$

Writing $\tau_\nu = \dfrac{3c^3 m}{8\pi^2 e^2 \nu^2}$ (so that τ_ν is the time, on the classical theory, for the energy of a point charge e and mass m performing simple harmonic oscillations of frequency ν to be reduced to $1/\epsilon$ of its original value, where ϵ is the base of logarithms), we have the expression

$$\frac{\mathbf{E}e^2}{4\pi^2 m}\left[\sum_a \left(\frac{a_a \tau_{\nu_a}}{\nu_a{}^2 - \nu^2}\right) - \sum_e \left(\frac{a_e \tau_{\nu_e}}{\nu_e{}^2 - \nu^2}\right)\right] e^{2\pi i\nu t},$$

or,

$$\frac{\mathbf{E}e^2}{4\pi^2 m}\left[\sum_a \left(\frac{f_a}{\nu_a{}^2 - \nu^2}\right) - \sum_e \left(\frac{f_e}{\nu_e{}^2 - \nu^2}\right)\right] e^{2\pi i\nu t} \quad \ldots\ldots(7),$$

where $f = a\tau = \dfrac{3c^3 ma}{8\pi^2 e^2 \nu^2}$ and may be called the 'strength' of the transition in question.

This dispersion formula (7) was given by Kramers in his first notice[1] on the quantum theory of dispersion, and in a second notice[2], the theory given in some detail above was briefly outlined.

This formula only contains such quantities as can be directly observed in connection with the spectrum of the atom concerned; all trace of the multiply periodic mechanism which represented the action of the atom has disappeared.

If the atom is in its 'normal' state then only the ν_a part of the formula is obtained (the first term), because no emission is possible; this form of the result had been found by Ladenburg[3] a few years before.

153. The terms of frequencies $\nu + \omega + \omega'$ in formula (4) have been translated into quantum theory terms by Kramers and Heisenberg[4], who have shown that these terms, which are not coherent with the incident light, have frequencies $\nu \pm \nu^*$, where $h\nu^*$ is the energy difference of the atom in the considered and any other state. The details of this beautiful and interesting work are however too long to be given here.

[1] H. A. Kramers, Nature, **113**, p. 673, 1924.
[2] H. A. Kramers, Nature, **114**, p. 310, 1924.
[3] R. Ladenburg, Zeitschr. für Phys. **4**, p. 451, 1921.
[4] H. A. Kramers and W. Heisenberg, Zeitschr. für Phys. **31**, p. 681, 1925.

To summarise: a stream of monochromatic light causes an atom to emit not only coherent waves of the same frequency as those of the incident light, but also non-coherent waves whose frequencies are combinations of that frequency with other frequencies corresponding with the possible transitions to other stationary states.

Thus a theory on which no *actual* transitions of the atom from one stationary state to another are involved, leads to the explanation of the scattered non-coherent waves in terms of *possible* transitions; it is a necessary part of this theory that such transitions should not occur, for if they did the energy condition (3) would no longer be valid.

Recent work on this subject is also due to Born[1] and Jordan[2].

154. These results as to the *frequency* of the scattered waves were first found by Smekal[3], who used the idea of 'light-quanta' to show that the frequencies $\nu \pm \nu^*$ must appear.

A light-quantum $h\nu$ is absorbed by the atom and another light-quantum $h\nu'$ given out. If the atom undergoes no transitional change, it acquires, if at rest before, a velocity v, where $h(\nu - \nu') = \frac{1}{2}mv^2$. Thus the change of frequency in the scattered rays is associated with a change of velocity corresponding to the Doppler effect of radiation from a moving source, observed by Compton[4] in his experiments with X-rays.

Smekal supposes that there must also be processes in the atom by which at the same time the atom undergoes a change to another stationary state, without change of velocity, so that if $h\nu_a$ is the quantum needed for the change to a state of higher energy $h\nu_a = h(\nu - \nu')$, and if $h\nu_e$ is the quantum given out in a change to a state of lower energy,

$$-h\nu_e = h(\nu - \nu').$$

Thus $\nu' = \nu - \nu_a$ or $\nu + \nu_e$ in the two cases.

[1] M. Born, Zeitschr. für Phys. **26**, p. 379, 1924.
[2] M. Born and P. Jordan, Zeitschr. für Phys. (to appear shortly).
[3] A. Smekal, Naturwissensch. **11**, p. 873, 1923.
[4] A. H. Compton, Phys. Rev. **21**, p. 483, 1923.

But, on the theory of Kramers and Heisenberg these results can be obtained by a generalisation of the classical theory and without the conflict with that theory involved in the use of light-quanta.

155. *The quantum-kinematics of Heisenberg.* It has been seen how the postulates of the quantum theory have led to an exact account of the properties of the hydrogen atom and of the effect of an electric or a magnetic field upon its spectrum.

But experience shows that in the problem of the 'crossed fields' (where hydrogen is exposed to the simultaneous action of an electric and a magnetic field whose directions are different) fundamental difficulties arise; further, that the reaction of the atom to periodic fields (as in the dispersion problem) cannot be described by these postulates alone; and lastly, that an extension of these quantum rules to the treatment of atoms with several electrons has so far proved itself impossible.

A new procedure, first used by Kramers in obtaining the quantum theory dispersion formula, endeavours to generalise a known formula of the classical theory, by the use of the correspondence principle, into a new formula of the quantum theory which asymptotically merges into the classical one for large quantum numbers.

A classical formula containing terms of the type $\mathbf{C}e^{2\pi i\omega t}$, where $\mathbf{C}$ is an amplitude vector and ω a frequency of the periodic system which represents the action of the atom, is translated into a quantum formula containing terms of the type $\mathbf{A}e^{2\pi i\nu t}$, where $\mathbf{A}$ is an amplitude vector directly connected with the probability of the transition which causes the frequency ν in the spectrum. The new formula is thus expressed in terms of the experimentally observable magnitudes of the quantum theory, namely, the intensities and frequencies of the spectral lines associated with the atom.

156. Heisenberg[1] has lately put forward the beginnings of a scheme of quantum-kinematics, which when more developed

[1] W. Heisenberg, Zeitschr. für Phys. 33, p. 879, 1925.

should lead to the direct deduction of these quantum theory formulae, without the intermediate use of the classical formulae in each problem considered.

In this first exposition of the new kinematics, Heisenberg considers only cases of one degree of freedom. A coordinate x is on the classical theory expressed as a Fourier series $\sum_\alpha \mathbf{C}_\alpha e^{2\pi i\omega\alpha t}$, where α is an integer and $\mathbf{C}$, ω are functions of the action variable I, which depends upon a number n.

Hence on the classical theory

$$x(n, t) = \sum_{-\infty}^{\infty}{}_\alpha \mathbf{A}_\alpha(n)\, e^{2\pi i\omega(n).\alpha t} \quad \ldots\ldots\ldots\ldots(1),$$

where $\mathbf{A}_\alpha$, ω are used as functional symbols.

On the quantum theory, he writes

$$x(n, t) = \sum_\alpha \mathbf{A}(n, n-\alpha)\, e^{2\pi i\omega(n, n-\alpha).t} \ldots\ldots\ldots(2),$$

where this $\mathbf{A}$ and this ω are the intensity and frequency corresponding to a transition $n \rightarrow (n-\alpha)$; this is Fourier's series generalised to a quantum form.

Also $$\omega(n, n-\alpha) = \frac{1}{h}[E(n) - E(n-\alpha)],$$

where E is the energy, and

$$\omega(n) = \frac{\partial E}{\partial I} = \frac{\partial E}{h\partial n},$$

so that for large quantum numbers

$$\omega(n, n-\alpha) = \frac{1}{h}\left(\alpha\frac{\partial E}{\partial n}\right) = \alpha\omega(n),$$

and the classical and quantum frequencies agree.

He then asks what is the formula for x^2 in the quantum-Fourier series?

The classical formula is seen to be

$$[x(n, t)]^2 = \sum_\alpha\sum_\beta \mathbf{A}_\alpha \mathbf{A}_{\beta-\alpha}\, e^{2\pi i\omega\beta t} \quad \ldots\ldots\ldots\ldots\ldots\ldots\ldots(3),$$

for this $$= \sum_\alpha\sum_\beta \mathbf{A}_\alpha \mathbf{A}_{\beta-\alpha}\, e^{2\pi i\omega(\alpha+\overline{\beta-\alpha})t}$$

$$= \sum_\alpha \mathbf{A}_\alpha e^{2\pi i\omega\alpha t} \sum_\beta \mathbf{A}_{\beta-\alpha}\, e^{2\pi i\omega(\beta-\alpha)t},$$

each being from $-\infty$ to ∞ for α and β and therefore for $\overline{\beta-\alpha}$,

$$= [x(n, t)]\,[x(n, t)].$$

This suggests the quantum theory formula

$$[x(n,t)]^2 = \sum_\alpha\sum_\beta \mathbf{A}(n, n-\alpha)\,\mathbf{A}(n-\alpha, n-\beta)\,e^{2\pi i\omega(n,n-\beta)t} \quad \ldots(4),$$

since $\quad \omega(n, n-\beta) = \omega(n, n-\alpha) - \omega(n-\alpha, n-\beta)$,

on account of the relation

$$\omega(n, n-\alpha) = \frac{1}{h}[E(n) - E(n-\alpha)].$$

And so on for powers of x, and therefore for any function of x which can be developed in powers of x.

157. The quantum condition $\oint p\,dq = nh$ has to be transformed to the new notation.

This condition is

$$nh = \oint m\dot{x}\,dx = \oint m\dot{x}^2\,dt.$$

In the classical notation

$$\dot{x} = \Sigma\,\mathbf{A}_\alpha\,2\pi i\omega\alpha\,e^{2\pi i\omega\alpha t},$$

so that $\dot{x}^2$, from (3),

$$= \Sigma\Sigma\,\mathbf{A}_\alpha\mathbf{A}_{\beta-\alpha}\,(2\pi i\omega\alpha)^2\,e^{2\pi i\omega\beta t},$$

$$\therefore\quad \oint m\dot{x}^2\,dt = m\int_0^{\frac{1}{\omega}} \Sigma\Sigma\,\mathbf{A}_\alpha\mathbf{A}_{\beta-\alpha}\,(2\pi i\omega\alpha)^2\,e^{2\pi i\omega\beta t}\,dt,$$

and the integrals on the right vanish for all the terms except those for which $\beta = 0$,

$$\therefore\ \oint m\dot{x}^2\,dt = m\Sigma\,\mathbf{A}_\alpha\mathbf{A}_{-\alpha}\,(2\pi i\alpha)^2\,\omega,$$

$$\therefore\quad nh = 4\pi^2 m\Sigma\,|\,\mathbf{A}_\alpha\,|^2\,\alpha^2\omega,$$

$$\therefore\quad h = 4\pi^2 m\Sigma\alpha^2\frac{d}{dn}[|\,\mathbf{A}_\alpha\,|^2\,\omega] = 4\pi^2 m\Sigma\alpha\frac{d}{dn}[\alpha\omega\,.\,|\,\mathbf{A}_\alpha\,|^2].$$

This suggests the corresponding quantum formula

$$h = 8\pi^2 m\Sigma\,[|\,\mathbf{A}(n, n+\alpha)\,|^2\,\omega(n, n+\alpha) - |\,\mathbf{A}(n, n-\alpha)\,|^2\,\omega(n, n-\alpha)] \quad \ldots\ldots(5).$$

Using these formulae (4), (5), Heisenberg works out the theory of the anharmonic oscillator $\ddot{x} + \omega_0{}^2 x + \lambda x^2 = 0$; and also the theory of the rotator, alone and with precession, confirming the results of Kratzer[1] for band spectra, and those of Goudsmit, Kronig, and Hönl[2] for the multiplets and for the intensities in the Zeeman effect.

[1] A. KRATZER, Sitzungsber. d. Bayr. Akad. p. 107, 1922.

[2] H. GOUDSMIT u. R. DE L. KRONIG, Naturw. **13**, p. 90, 1925; H. HÖNL, Zeitschr. für Phys. **31**, p. 340, 1925; R. DE L. KRONIG, Zeitschr. für Phys. **31**, p. 885, 1925; A. SOMMERFELD u. H. HÖNL, Sitzungsber. d. Preuss. Akad. p. 141, 1925.

INDEX OF AUTHORS

The numbers refer to the pages

www.ingramcontent.com/pod-product-compliance
Ingram Content Group UK Ltd.
Pitfield, Milton Keynes, MK11 3LW, UK
UKHW040704180125
453697UK00010B/407